Springer Series in Solid-State Sciences

Volume 203

The Springer Series in Solid-State Sciences features fundamental scientific books prepared by leading and up-and-coming researchers in the field. They strive to communicate, in a systematic and comprehensive way, the basic principles as well as new developments in theoretical and experimental solid-state science.

We welcome submissions for monographs or edited volumes from scholars across this broad domain. Topics of current interest include, but are not limited to:

- Semiconductors and superconductors
- Quantum phenomena
- Spin physics
- Topological insulators
- Multiferroics
- Nano-optics and nanophotonics
- Correlated electron systems and strongly correlated materials
- Vibrational and electronic properties of solids
- Spectroscopy and magnetic resonance.

Navin Khaneja

Electron Waves in Solids

Exploring the Foundations of Modern Technology

Springer

Navin Khaneja
System and Control Engineering
Indian Institute of Technology Bombay
Mumbai, Maharashtra, India

ISSN 0171-1873 ISSN 2197-4179 (electronic)
Springer Series in Solid-State Sciences
ISBN 978-3-031-67259-0 ISBN 978-3-031-67260-6 (eBook)
https://doi.org/10.1007/978-3-031-67260-6

This Springer imprint is published by the registered company Springer Nature Switzerland AG
The registered company address is: Gewerbestrasse 11, 6330 Cham, Switzerland

Dedicated to Darshan Lal Khaneja

Preface

The origin of this book can be traced back to the author's research in nuclear magnetic resonance (NMR) spectroscopy. Modern NMR spectrometers use a superconducting magnet to create a very large and stable magnetic field. The superconducting coil in a magnet carries a current that never dissipates. One just has to cool the coil with liquid Helium and it operates round the clock. What an amazing physical phenomenon, called superconductivity. Understanding superconductivity has led the author to dig into the beautiful subject of solid state physics. How are electrons in a solid? Are they waves or particles? Why are certain materials conductors while other insulators? How do we understand modern electronic devices ranging from diodes and transistors to MOSFET, LEDs, and solar cells. The book tries to do justice to all this. In writing the book, I have tried to answer all the questions, I had as an electrical engineering undergraduate at IIT Kanpur, taking a device physics course. I have chosen subjects that I feel are conceptual and challenge out ability to visualize how things are. I cover traditional topics like band theory of solids and phonons to more application oriented topics like electronic devices and exciting research areas like superconductivity and quantum hall effect. The book is written as a research monograph, but it is more an effort on the author's part to sow the world of solid state, to paint a bigger picture, place where things are, and assimilate various concepts. The book is intended for a first course in solid state physics or condensed matter theory. The focus is more on a bigger picture, which can be supplanted with exercises for which there are many excellent textbooks around. It is the author's hope that the book will come in handy for researchers in the broad area of solid state physics.

It is an opportunity to acknowledge numerous people who I believe directly and indirectly helped with this effort. I would like to thank Prof. Steffen Glaser and Niels Nielsen for numerous years of excellent collaboration in NMR spectroscopy that ultimately led me to develop this text. I would like to thank Professor Roger Brockett, who helped me in nurturing my taste in physics. I am grateful to Professor Sumiran Pujari and Soumya Bera at IIT Bombay for their informative lectures in condensed matter that helped refine my outlook toward the subject. I would like to thank the wonderful colleagues and academic environment of SYSCON at IIT Bombay that provided ample opportunity for self development. Finally, I like to acknowledge the support of my family which made this effort possible.

Mumbai, India
December 2018

Navin Khaneja

Contents

Part I
Electrons in Solids

How are electrons in a solid say a metal? High school text say they are mobile electrons, free to move under an applied electric field. What does a solid state physics text say? Electrons are waves. Take a block of metal say a feet long. Then one is made to believe it is filled with electron waves. The waves are characterized by their dispersion relation, which relates its time frequency ω to spatial frequency k. Waves have different spatial frequencies, and each k is occupied by two electrons with opposite spin. How do you reconcile the two pictures? A natural approach is to model electrons as wavepackets. Wavepacket is a superposition of a band of spatial frequencies which localizes it to a volume and it moves like a classical particle. If the band is centered at the spatial frequency k_0 then $\frac{d\omega}{dk}|_{k_0}$ is its velocity. Thus we get wavepackets of different velocities in a local volume, all of them replicated at different spatial locations/volumes filling the whole volume of solid. This is our picture of mobile electrons. The different velocity wavepackets in a local volume constantly move to adjacent volumes. When we apply an electric field say along x direction, the wavepackets accelerate in that direction. This is electric current. Electrons are moving at very high velocity up to 10^5 m/s (fermi velocity) but this motion averaged over all the electrons in volume gives no net momentum. The current arises due to additional velocity imparted due to the applied electric field which adds us over a volume. This is much slower at say drift velocity of 10^{-3} m/s for an ampere current through a wire of cross part 1 mm^2. What limits this drift velocity? On application of an electric field, electrons accelerate and gain momentum but collide with a phonon packet, a localized potential arising due to deformation of the lattice. This collision rebounds the electrons, breaks its velocity, and limits the drift velocity. This is called resistance. If τ is the average time between collisions, so-called relaxation time, then the velocity acquired between collisions is $\frac{eE\tau}{m}$, which is the mean drift velocity.

It is not always possible to accelerate. What does acceleration mean when electrons are wavepackets. One moves to a higher k-state (meaning higher velocity). When all k-states are occupied, as in an insulator, we cannot accelerate, we cannot have current. In a conductor, k-sates are partially filled and it is possible to accelerate. Collision with the phonon exchanges energy between electrons and phonons. This gives a distribution on the possible velocities of wavepackets in a local volume. This is called Fermi–Dirac distribution.

We have been talking about electron wavepackets and how they collide with phonons to give resistivity in metals. There is another very interesting phenomenon that takes place in solid state physics when certain metals are cooled below the critical temperature of the order of a few kelvin. The resistance of these metals completely disappears and they become superconducting. The electrons in a local volume bind with phonon mediated interaction. Collisions with a phonon will rebound an electron and break this bond. The phonon does not have enough energy at low temperatures to break this bond hence electrons are not scattered during phonon collisions at very low temperatures of a few kelvin, this is superconductivity.

Chapter 1
Localized Electron Waves

1.1 Introduction

What is solid state physics [1–8]? It is not easy to give a simple and complete answer, but let's try. Take a solid material like a metal, which conducts electricity. The metal has atoms arranged in a periodic fashion. We are taught in high school physics, that there are mobile electrons in the metal. How do they move when, say I apply an electric field. Closer inspection will tell answering this is a daunting task. After all, there is a forest of atoms so-called lattice, which will attract (pull) the electron. We can write the electric field the electron sees due to all of the lattice and the electric field that we apply and try to solve for the Newton equations describing the motion of an electron. Not only is this a difficult task, as one might suspect, but the electron will just end up banging against the lattice. But what if I say, if everything is done correctly, then all you need to do is to just change the mass of the electron and simply ignore all of the lattice. The electron just moves under the electric field as if it were all free with a modified mass. That is the simplicity reached by the subject of solid state physics. Therefore the picture developed in high school of mobile carriers, free to move under an electric field is made all true in solid state physics by changing the electron mass.

But now your electron is not like a marble. Rather it is a localized wave. A lump of energy, with spatially and time varying phase. You may say, it is like a cloud moving through the lattice. The clouds can occupy the same volume as other clouds (something not true of marbles), as long as their spatial frequencies are distinct. Different spatial frequencies mean different velocities, different masses, and different energies.

These electron clouds come from atoms that form the solid. When these atoms are isolated, the electrons live in them as orbitals. When they are all brought together to form a solid, the electrons that had a home in the atoms end up as electron waves. The electrons from different atoms form waves, that occupy the same volume, except now they have different spatial wave frequencies and energies. In fact, these waves

N. Khaneja, *Electron Waves in Solids*, Springer Series in Solid-State Sciences 203,
https://doi.org/10.1007/978-3-031-67260-6_1

Fig. 1.1 **a** Depicts dispersion curve for a free electron. **b** Depicts a typical dispersion curve for a solid

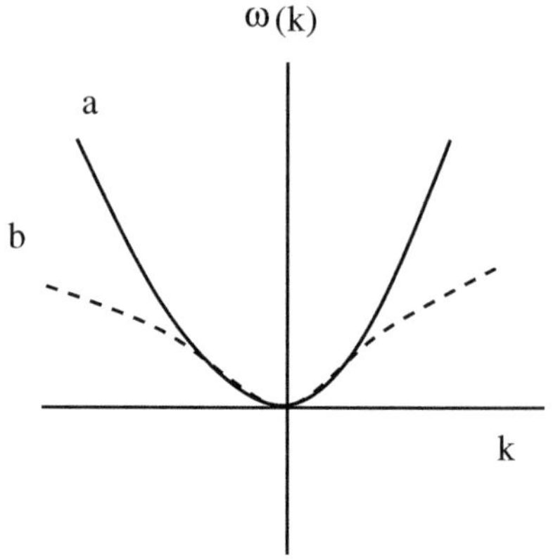

form a band of energy. Waves from different orbitals are different and may form distinct energy bands with typical gaps in their energy spectrum.

Now we can answer what solid state physics is. It is the study of these electron waves. Solid state physics is about finding the energies of these waves and their dependence on their spatial frequencies so-called dispersion relations. If ω is the frequency of the wave, and k its spatial frequency, then from quantum mechanics, the energy and momentum is $E = \hbar\omega$ and $p = \hbar k$. The relation $E = \frac{p^2}{2m}$ relates ω to k, written as $\omega(k) = \frac{\hbar k^2}{2m}$. But this is a free electron. For an electron wave in a lattice, we get a different dispersion relation $\omega(k)$, which is what solid state physics finds. Figure 1.1a depicts the dispersion curve for a free electron. Figure 1.1b depicts a typical dispersion curve for a solid.

The electron waves are delocalized. If we have a conductor of length $L = 1$ m, we have these waves over a length of 1 m. This is a bit counterintuitive. Our high school picture is of mobile point-like particles, which we understood as a moving cloud, a localized wave. How do we go from wave to localized wave? This means the wave not only has a spatial frequency k but also a bandwidth of spatial frequencies Δk. When we sum waves with frequencies in this bandwidth, we get a wavepacket that is localized. Figure 1.2A depicts a k-point dispersion curve corresponding to a wave. Figure 1.2B depicts that when there is the width to k-point, it corresponds to a wavepacket of finite width and there are replicas of such packets, spatially displaced. There are as many replicas as number of k-points in the bandwidth. Is there a natural bandwidth Δk and resulting localization volume? More is said about this in the subsequent section, but now we are closer to the picture of an electron particle in terms of an electron cloud, which we also call localized electron wave, wavepacket, or quasiparticle. Figure 1.3 depicts such an electron quasiparticle, which is in fact a localized wave and moves with a modified mass.

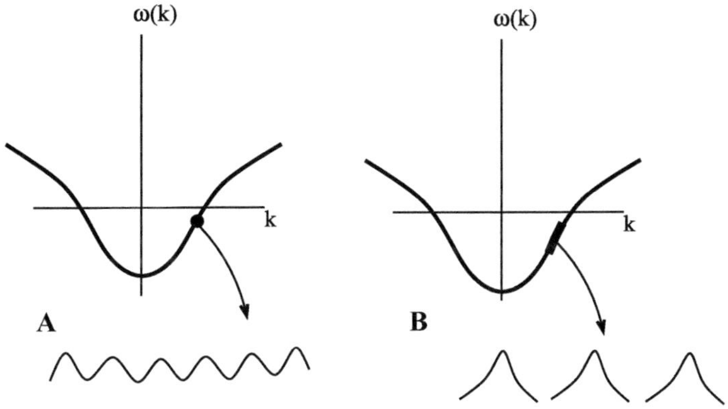

Fig. 1.2 **A** Depicts a k-point on the dispersion curve corresponding to a wave. **B** Depicts, when there is the width to the k-point, it corresponds to a wavepacket of finite width, and there are replicas of such packets spatially displaced

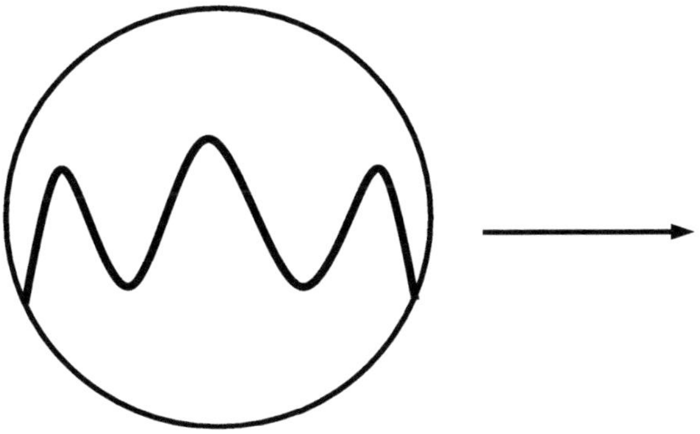

Fig. 1.3 Figure depicts an electron quasiparticle. A particle, which is in fact a localized wave, moving with a modified mass

If wavepacket is centered at spatial frequency k_0, its velocity $v_g = \frac{d\omega}{dk}|_{k_0}$. If we apply an electric field E, as we show in subsequent section, we evolve spatial frequency $k_0 \to k_0 - eEt/\hbar$, and hence $\frac{dv_g}{dt} = \underbrace{(\hbar^{-1}\frac{d^2\omega}{dk^2}|_{k_0})}_{1/m^*} eE$, where m^* is the modified mass. Figure 1.4A depicts a dispersion curve in a solid, with electron wavepackets (made of a bunch of k states) shown in thick. We see wavepackets with forward and backward velocity. Figure 1.4B shows what happens when we apply an electrical field (pointing left), the k values change and we get a wavepacket with higher k. So that is all. In solids, electrons are like wavepackets with modified mass moving like free mobile particles. If that is all, then why are some solids good conductors while others are insulators and we also have materials called semiconductors, which conduct at high temperatures and not at low temperatures. How do we understand conductivity in terms of electron waves?

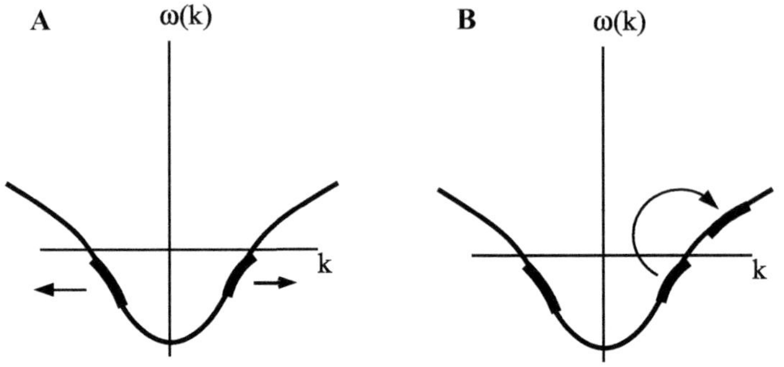

Fig. 1.4 **A** Depicts dispersion curve in a solid, with electron packets (made of a bunch of k states) shown in thick. We see wavepackets with forward and backward velocity. **B** Shows what happens, when we apply an electrical field (pointing left), the k values changes, and we get wavepacket with higher k

If there are n atoms, with one electron per atom in a given orbital, then after the solid is formed, we have n electron waves with different spatial frequencies. If we think of a solid in one dimension with atoms a units apart, then these spatial frequencies vary between $[-\frac{\pi}{a}, \frac{\pi}{a}]$ (we will see this later), say equally spaced. We can put one electron per frequency and fill all frequencies or we can have 2 electrons per frequency and only occupy $\frac{n}{2}$ frequencies say the ones with lower energy. This is called half filled band. This is shown in Fig. 1.5A, where a typical dispersion curve $\omega(k)$ is shown, and filled k states are shown in dark. The k states have both left and right velocities, and the net momentum is zero. When we apply an electrical field (say pointing left), the spatial frequencies evolve and we get a configuration as in Fig. 1.5A with more electrons with right velocity and hence we have net current from left to right. This is a conductor.

Now if there are n atoms with two electrons per atom in a given orbital, then after the solid is formed we have n electron waves all filled and we have fully filled band as shown in Fig. 1.5C. When we apply an electric field, nothing happens as k evolves from left to right cyclically, resulting in no change of net momentum and hence no net current. We have what we call an insulator. A material with filled bands. But it is possible that there may be another band (arising from a different orbital) that is empty. This band may have energies not very high compared to the filled band as shown in Fig. 1.5D. Then due to finite temperatures, there is a finite probability that instead of occupying all low energy band, we occupy some of these high energy states. This higher energy band is not filled and we can conduct. This is called a semiconductor, where at high temperatures occupancy in unfilled higher energy band increases and we can conduct.

Why do electron orbitals become electron waves? Remember electron orbitals are also waves, but highly localized. They come about when we solve Schroëdinger equation of electron in the central potential of an ion. Now in a solid, our potential is different. It is the potential of all the nuclei in a solid. We have to solve for electron

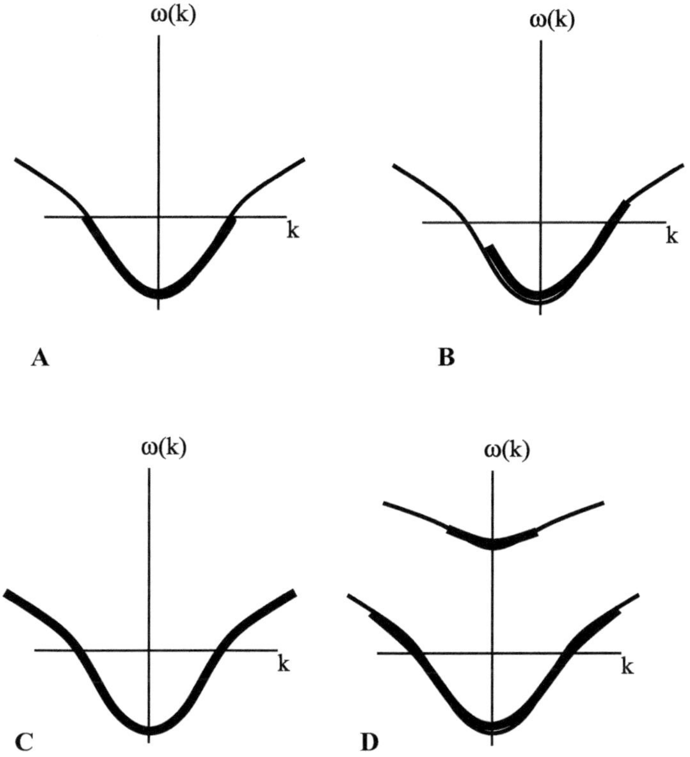

Fig. 1.5 **A** Depicts half filled dispersion curve in a solid. **B** Shows what happens, when we apply an electrical field (pointing left), the half filled band develops a net right momentum

orbitals in this new potential. Electron waves are precisely these orbitals. To fix ideas, think of the linear array of atoms in 1D with a as atomic separation. The potential is a periodic potential, which is replicated at all atomic sites. We first solve for Schroëdinger equation in this rapidly varying periodic potential. We get electron waves as stationary states with a dispersion relation $\omega(k)$.

How to solve for Schroëdinger equation of electron in a periodic potential. There are two ways to proceed. The first one is called nearly free electron approximation. The periodic potential with periods a may be expanded by Fourier series as the sum of exponentials $\exp(i\frac{2\pi mx}{a})$. We may start with free electron waves $\exp(ikx)$, which in the absence of any potential has kinetic energy $\frac{\hbar^2 k^2}{2m}$. Such free waves, whose wavevectors differ by $\frac{2\pi}{a}$ are connected (mixed) by the potential, then we get a linear combinations $\exp(ikx)\underbrace{\{1+\sum_m b_m \exp(i\frac{2\pi mx}{a})\}}_{p_n(x)}$ as eigenvectors to this

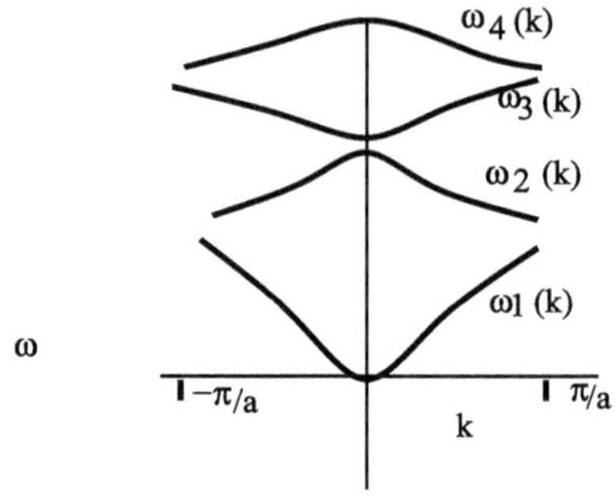

Fig. 1.6 Figure depicts energy bands

problem with energy $\epsilon_n(k) = \hbar\omega_n(k)$, with energy increasing as n. This is called a Bloch wave, with spatial frequency k and $p_n(x)$ periodic with period a. Clearly by construction we can curtail $k \in [-\frac{\pi}{a}, \frac{\pi}{a}]$. We can plot different energy bands $\omega_1(k), \omega_2(k), \ldots$ as in Fig. 1.6.

In the above approach, for solving the Schroëdinger equation in a periodic potential, we started with free electron waves. There is another approach called tight-binding approximation, where we start with orbitals ϕ_i with energy ϵ_0 on atomic sites. When atoms are brought into the vicinity, these orbitals overlap and feel the potential of neighboring ions. Potential V_{i+1} at site $i+1$ scatters ϕ_i to ϕ_{i+1}, and we define a scattering (transfer) term $-t = \langle \phi_{i+1} | V_{i+1} | \phi_i \rangle$. Then starting from ϕ_i we can form new eigenfunctions of the Hamiltonian

$$H = \begin{bmatrix} \epsilon_0 & -t & 0 & \ldots & \ldots & -t \\ -t & \epsilon_0 & -t & 0 & \ddots & \vdots \\ 0 & -t & \epsilon_0 & -t & 0 & \vdots \\ \vdots & \ddots & \ddots & \ddots & \ddots & \vdots \\ 0 & \ldots & \ldots & -t & \epsilon_0 & -t \\ -t & 0 & \ldots & \ldots & -t & \epsilon_0 \end{bmatrix}, \tag{1.1}$$

where for mathematical convenience, we assume the linear array is wound up as a circle. The eigenvalues are $\epsilon_0 - 2t \cos ka$ with eigenfunction $(1, \exp(ika), \ldots, \exp(imka), \ldots)$, with $k \in [-\frac{\pi}{a}, \frac{\pi}{a}]$, equally spaced at $\Delta k = \frac{2\pi}{na}$. This is a wave solution, with spatial frequency k. We start with orbitals, and as we bring them together, they get mixed to form waves with energy ϵ_0, spreading into a band $\epsilon_0 - 2t \cos ka$, as shown in Fig. 1.7. The dispersion relation is as shown in Fig. 4.14. Observing in the nearly free electron method, we get many bands as shown in Fig. 1.6. In tight-binding approximation, also we will get many bands, when we start with different energy orbitals, with each getting spread into a band as shown in Fig. 1.9. The inner orbitals don't overlap much, and hence their t is small, and the spread of the band which is $4t$ is small, as opposed to outer orbitals, with large t, giving bands

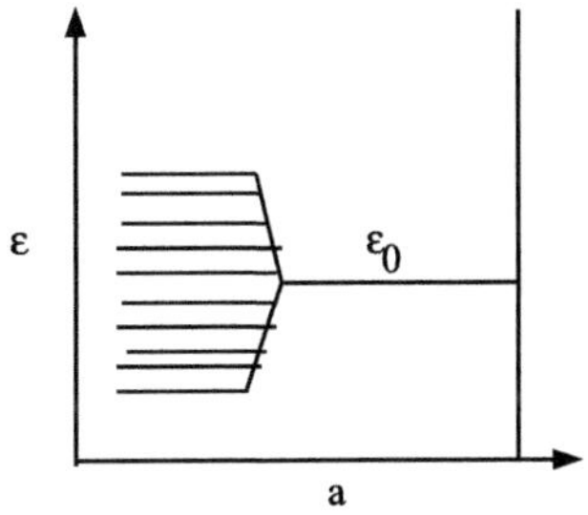

Fig. 1.7 Figure shows how atomic orbital energy ϵ_0 gets branched into many energies as we decrease the separation a between orbitals

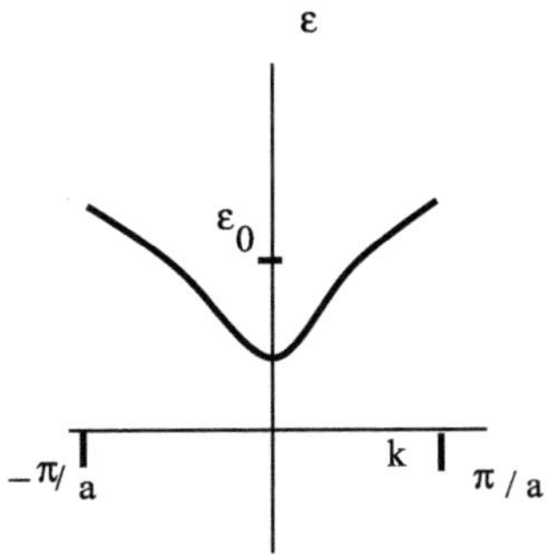

Fig. 1.8 Figure shows a plot of dispersion relation in (4.38)

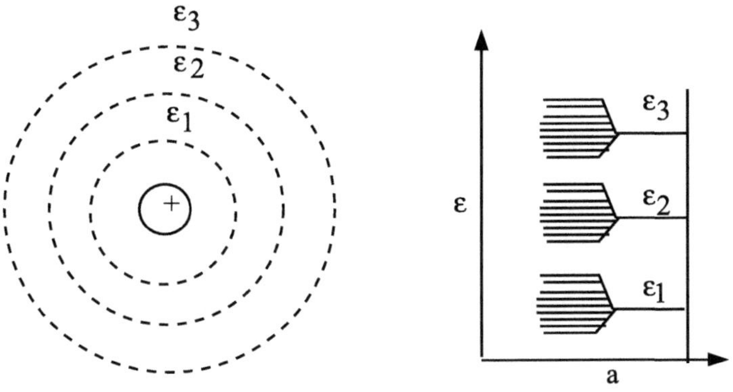

Fig. 1.9 Figure shows how many orbitals with energies $\epsilon_1, \ldots, \epsilon_N$ get broadened as we decrease the separation a between orbitals

with larger bandwidths. Outer s orbitals give bands with bandwidths of order 5−10 eV while d orbitals are localized and give bands that are narrow of order 1−2 eV (Fig. 1.8).

We conclude this section by saying our electrons are localized electron waves or wavepackets moving around in a conductor. We say we have a Fermi gas of electrons as shown in Fig. 1.10.

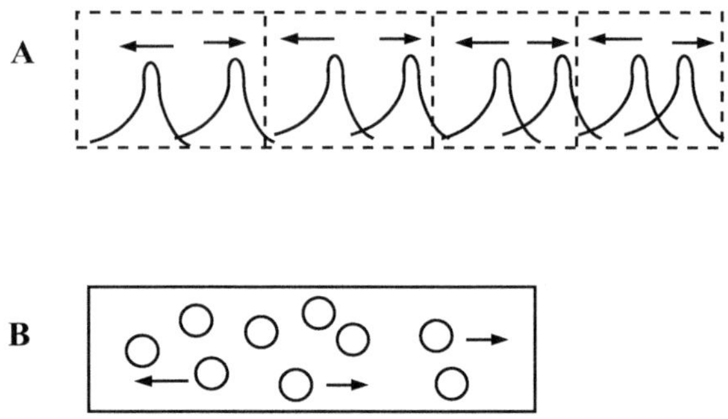

Fig. 1.10 **A** Shows localized packets moving from one local volume to another at Fermi velocity. **B** Depicts Fermi gas of electrons in a metal moving in a conductor

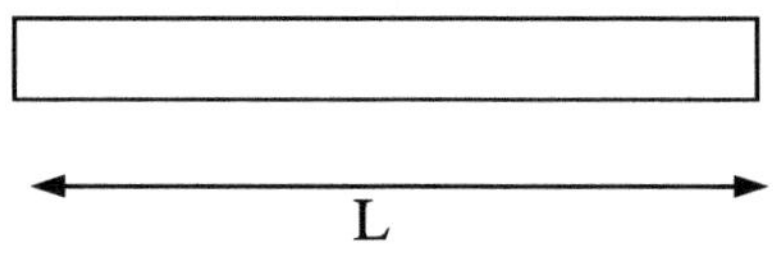

Fig. 1.11 Figure depicts a metallic bar

1.1.1 Particles and Waves

Shown in Fig. 1.11 is a bar of metal. How are electrons in this metal bar? At the outset, we may distinguish between electrons that are bound to the atoms (those in the inner electronic shells) and the mobile electrons (that come from the outermost electronic shells). As taught in high school physics, these mobile electrons are shared among all the atoms and are free to move around. Then at most elementary levels, we can think of these electrons as hard spheres, moving in the conductor very much like gas molecules in a container as in Fig. 1.12A. But this description neglects the wave nature of these electrons.

Quantum mechanics tells us that electrons are waves with a wavefunction. Then may be the correct picture of mobile electrons is Fig. 1.12B. The electron wave $\psi = \exp(ikx)$ fills the whole conductor and for it to be localized to conductor length $L = na$ it must have a bandwidth (it is a wavepacket) of $\Delta k = \frac{2\pi}{L}$. Then we can have wavepackets centered at $k = 0, \pm 2\Delta k, \ldots, \pm n\Delta k$. These electron waves (a typical one is shown in Fig. 1.12B.) all fill the whole conductor and have non-overlapping spatial frequencies (different k values) and hence are orthogonal. This is the usual picture developed in a course in solid state physics. However, there is a third possibility, which is intermediate between the above two as shown in Fig. 1.12C. Electron waves are now localized in the conductor to length L_0. They are localized wavepackets. These wavepackets must have much larger bandwidth $\Delta k_0 = \frac{2\pi}{L_0}$. Wavepackets centered at $0, \pm 2\Delta k_0, \ldots, \pm n\Delta k_0$, occupy the same region of the conductor and these same spatial frequencies again repeat themselves in different non-overlapping regions. This is the picture we develop in this book, the picture of localized waves.

Let's say our metallic bar is $L = 30$ cm long with inter-atomic spacing of $a = 3$ A°. Then there are 10^9 atoms in the metal bar, say each giving a mobile electron. Then in a classical approach, we can have 10^9 electron waves spaced very finely (see

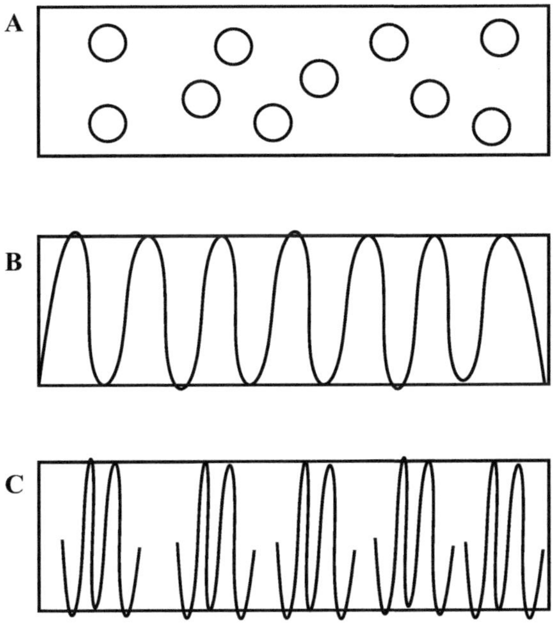

Fig. 1.12 Figure depicts a metallic bar

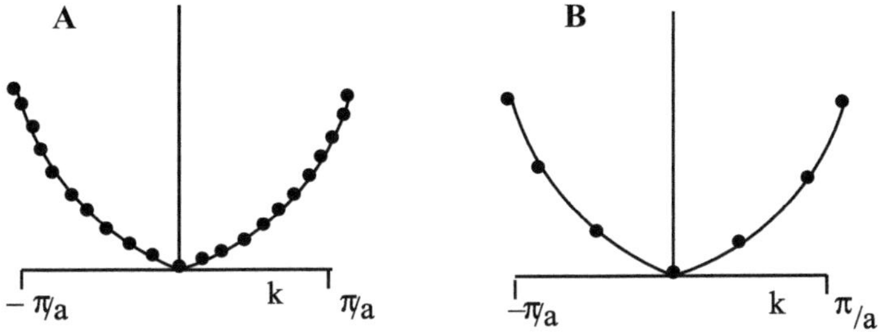

Fig. 1.13 Figure shows dispersion relation for electron wave (**A**) and localized electron waves (**B**). The former are finely spaced in the k domain while later one are much coarsely spaced

Fig. 1.13A) at $\Delta k = 10^{-9}\,\frac{2\pi}{a}$, each holding an electron (actually each can hold two electrons so we really only need half as many). These waves fill the whole conductor length L. Now instead, if we consider electron waves confined to $L_0 = 300$ A°. Then there are only 100 odd atoms in this length and we only have 100 odd localized waves in this region, with a coarse spacing (see Fig. 1.13B) $\Delta k_0 = \frac{2\pi \times 10^{-2}}{a}$. The replica of these waves occupies the non-overlapping region of length L_0. Since there are 10^7 such non-overlapping regions, if we count all the waves, they again turn out to be 10^9, the same number just more localized, with coarser spread in their spatial frequency. They are particle-like because of localization yet they are a wave with a spatial frequency.

We talked about electron waves with their energies organized into bands. Let's take metal sodium with electronic configuration $1s^2 2s^2 2p^6 3s^1$. Orbitals 1s, 2s, and 2p form fully filled bands as the respective orbitals are filled. The orbital $3s$ forms a half filled band. Half filled band means lower half (in energy) of k states are occupied with two electrons each. But that is literary true only at zero temperature. At room temperature, there is a nonzero probability of occupancy of higher k-states. This probability so-called Fermi–Dirac distribution is given by

$$P(\epsilon) = \frac{1}{1 + \exp\left(\frac{(\epsilon - \mu)}{kT}\right)},$$

where ϵ is the energy of k-state and μ the chemical potential and T temperature. How does this probability come about? Electrons exchange energy with heat bath which in this case is the lattice vibrations. At finite temperature, atoms are not stationary, but oscillate giving rise to lattice wave phonons. We talked about electron waves and resulting wavepackets. Similarly, we have phonons and phonon packets. These packets of lattice deformation travel like classical particles at the speed of sound. Electron packets may collide with phonon particles, like a collision between two classical particles, which leads to an exchange of energy between them. The electron may lose or gain energy at the expense of the phonon. This way electron-like gas molecules talks to bath and exchange energy. Now there are constraints on electron velocity. Not all electrons can have the same velocity (or k values) as they are fermions. This gives rise to a distribution of electron velocity (energy) called Fermi–Dirac distribution.

When electron wavepackets collide with phonon packets, they exchange momentum and energy. After a collision, the electron will emerge out in a random direction depending on the direction of the advent of the phonon packet. Average momentum after collision is 0. If τ is the inter-collision time, then with probability $(1 - \frac{dt}{\tau})$ there is no collision in time dt. Then initial momentum $p(t)$ in time dt becomes

$$p(t + dt) = (p(t) - eEdt)\left(1 - \frac{dt}{\tau}\right), \tag{1.2}$$

which gives

$$\frac{dp}{dt} = -\frac{p(t)}{\tau} - eE. \tag{1.3}$$

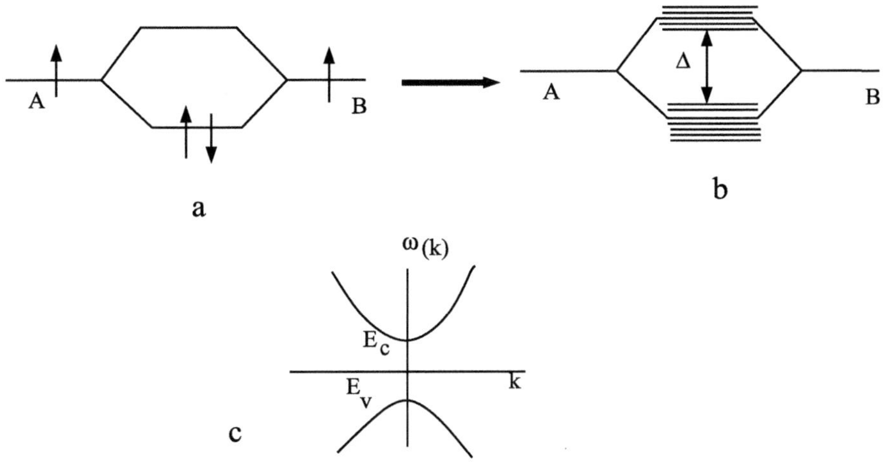

Fig. 1.14 Figure shows molecular orbitals are formed from atomic orbitals as in (**a**), which due to overlap between them get further broadened as in (**b**). **c** Shows conduction and valence bands. E_c is energy at bottom of the conduction band and E_v at the top of valence

Electric field accelerates electrons and generates momentum while collisions ruin it. Thus we get a steady state when $p = mv = eE\tau$. The current density $j = nev$ where n is the electron density, which gives $j = \underbrace{\frac{ne^2\tau}{m}}_{\sigma} E$, where σ is the conductivity or $\rho = \sigma^{-1}$ the resistivity. This is Drude's model of conductivity.

As discussed before in semiconductors, due to Fermi–Dirac distribution, we get electrons into higher lying bands (conduction band) which can conduct, while lower filled ones (valence band) don't. Let's take the most important semiconductor silicon, with electronic configuration $1s^2, 2s^2, 2p^6, 3s^2, 3p^2$. The inner orbitals form a closed band. The outer s and p orbitals combine to form four sp^3 orbitals. The sp^3 orbitals between neighboring silicon bond to form a bonding orbital and antibonding orbital with an energy gap of 1–2 eV. These orbitals have further overlap along them which spreads them into bands. This is as shown in Fig. 1.14. The lower band arising from bonding orbital is the valence band and the higher one, we call the conduction band. At low temperatures, the valence band is full and the conduction band empty, but as we raise the temperature, we get electrons into the conduction band.

Temperature is not the only way to get electrons into the conduction band. Another way is chemically, by act of doping, where we substitute some silicon atoms with phosphorus. Phosphorus with electronic configuration $1s^2, 2s^2, 2p^6, 3s^2, 3p^3$, has extra electron in the outer shell compared to silicon. After band formation, this extra electron goes to the conduction band. This is called n-doped silicon and acts as conductor with current in the conduction band. If we substitute some silicon atoms with aluminum. Aluminum with electronic configuration $1s^2, 2s^2, 2p^6, 3s^2, 3p^1$ has one less electron, in the outer shell compared to silicon. Hence valence band is not

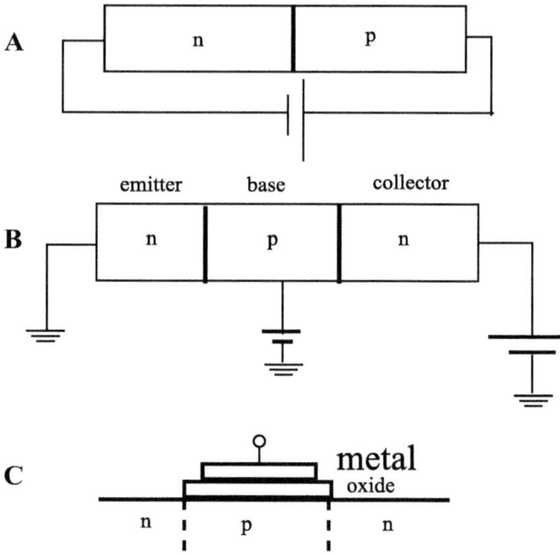

Fig. 1.15 Figure shows various devices made out of doped silicon

completely full and it can conduct. One way to see this is, that the valence band has missing wavepackets. Then when we apply the electric field, the filled band doesn't generate any net momentum or current then the current is in the opposite direction to that due to only the missing wavelet. This is called p-doped silicon, which conducts in the valence band. Both n- and p-doped silicon conduct at room temperature.

Interesting electronic devices can be made out of n- and p-doped silicon. For example, if we take a n-doped silicon bar, and form a junction with p-doped silicon bar as shown in Fig. 1.15A, we get a device that only conducts in one direction. If we apply a positive voltage to p-region, it pulls valence electrons, creating a positive void, which is filled by the movement of conduction electrons from n-region to p-region. The valence electrons in n-region don't come in to fill the void, because they cannot be accelerated, they are a filled band. However, it is different when we apply a positive voltage to n-region. We won't be able to pull valence electrons in n-region as they are not to be accelerated, nor can we move the conduction electrons, because who will take their place? There are no conduction electrons in the p-region. The valence of p can come in but where will they sit, all the valence bands in the n-region are full. This means no current if we pull from the p-side. Conduction happens only one way. The device is called a diode. It is used for rectifying voltages, as only conducts when p is positive compared to n and not the other way.

Another useful device one can make out of doped silicon is a so-called bipolar transistor. Consider a p-doped silicon bar (called the base) sandwiched between two n-doped silicon bars (the left one called emitter and the right one collector) as shown in Fig. 1.15B. When we apply a positive voltage on the base compared to the emitter, we pull valence electrons from the base, that are replaced by conduction electrons from the emitter. As these conduction electrons from the emitter enter the base, we pull them to the collector side by applying a more positive voltage on the collector compared to the base. Most of them (say fraction $1 - \alpha$) are pulled to collector only a fraction α go down from conduction to valence band in the base region and form the base current. This is called recombination. The ratio $\beta = \frac{1-\alpha}{\alpha}$ of the collector current to base current is called the current gain of the transistor. β is high of order of 100–1000 or more. Thus we have a device, where we modulate the base current by changing the base voltage and hence modulate the collector current with base voltage. We can then build amplifiers, etc.

The third useful device one can make out of doped silicon is a so-called MOSFET transistor as shown in Fig. 1.15C. We have n-doped regions, called source, and drain with a p channel between them. The channel has no conduction electrons and hence doesn't conduct, if say, I pull on the drain with positive voltage. But if we apply a positive voltage on the channel, with a metal electrode, which pulls electrons to the top, as shown in Fig. 1.15C. These extra electrons go in the conduction band and form so-called a thin inversion layer of conduction electrons, between drain and source. This then conducts. By using a control voltage on the gate, we can make the transistor to switch on and off. MOSFET forms the heart of most modern digital electronics.

We talked about silicon, which is an indirect bandgap material, as shown in Fig. 1.16A. The bottom of the conduction band is at a different k value than the top of the valence band. However, if we take a semiconductor like Gallium Arsenide (GaAS), we have a direct bandgap material as shown in Fig. 1.16B, where the bottom of the conduction band is at the same k value than the top of the valence band. When an electron falls from conduction to valence band, the difference of energy ΔE which is in 1–2 eV may be given out as light with frequency $\hbar\omega = \Delta E$. This ω is in optical range 10^{15} Hz. The momentum of light (its wavenumber) is ω/c where c is the velocity of light is much smaller than the electron momentum $\frac{\pi}{a}$, therefore if momentum is to be conserved, we cannot change very much the momentum of electron when it falls. Therefore, the mechanism of energy release using light is only possible in a direct bandgap material. Now we can make our p-n junctions using direct bandgap material like GaAs and pull electrons from the n-side conduction band to p-side and make them fall into the valence band, releasing light. This way we can make light emitting diodes. The frequency of light depends on the bandgap, for example in GaAs, we get a red light, in GaP we get the green light and in GAN we get blue light as the bandgap increases from red to blue. If the emitted light is

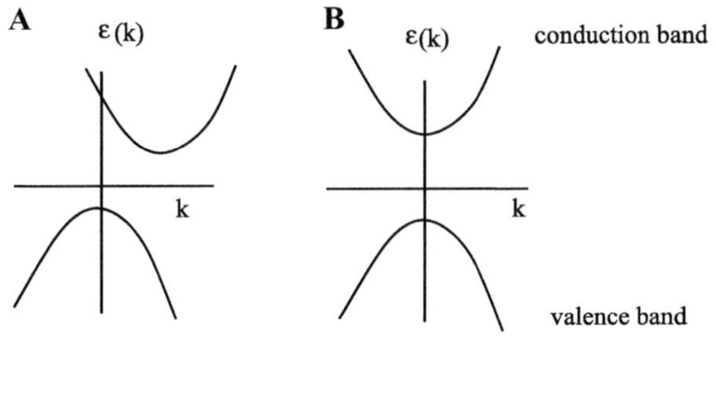

Fig. 1.16 **A** Shows indirect bandgap material. **B** Shows direct bandgap material

collected in a cavity, and made to travel back and forth, it induces (stimulates) more emissions, which have the same wavevector and phase giving highly directional light with the same phase. This is a semiconductor laser. LED and lasers are the heart of modern optical communication.

We have been talking about electron wavepackets and how they collide with phonons, to give resistivity in metals. There is another very interesting phenomenon that takes place in solid state physics, when certain metals are cooled below the critical temperature of order of few kelvin. The resistance of these metals completely disappears, and they become superconducting. How does this happen? One may guess, maybe at low temperatures there are no phonons. That is not true, as we have low frequency phonons present. Why do we then lose all resistivity? Electrons can bind together to form a molecule by phonon mediated interaction. The electron can pull on the lattice, which pulls on another electron. This phonon mediated bond is not very strong, only few meV, but at low temperatures, this is good enough, we cannot break it with collisions with phonons, which only carry $k_B T$ amount of energy, which is small at low temperatures. Then electrons cannot scatter them with phonon collisions.

We like to mention another fundamental solid state phenomenon, the phenomenon of magnetism. Magnetism has many manifestations, like ferromagnetism, paramagnetism, diamagnetism, antiferromagnetism, etc. Ferromagnetism is the basic mechanism by which certain materials (such as iron) form permanent magnets, or are attracted to magnets. An everyday example of ferromagnetism is a refrigerator magnet used to hold notes on a refrigerator door. Permanent magnets are either ferromagnetic. Only a few substances are ferromagnetic. The common ones are iron, nickel, cobalt and most of their alloys, and some compounds of rare earth metals. Ferromagnetism is very important in industry and modern technology and is the basis for many electrical and electromechanical devices such as electromagnets, electric motors, generators, transformers, and magnetic storage such as tape recorders, and hard disks, and nondestructive testing of ferrous materials.

How to understand ferromagnetism. We have half filled conduction band in metals, with two electrons per k-state spin up and down. In ferromagnetic materials, the band we care about is made out of d-orbitals, d-band. These bands are narrow in bandwidth, meaning electrons have small kinetic energy and hence they give localized electrons. By localizing electrons to atomic sites, we minimize repulsion between them. In ferromagnetism, there are more localized up spins than down. This is because when the spins on neighboring sites are both in the upstate, their spatial wavefunction should be in an anti-symmetric state so that the overall wavefunction is anti-symmetric. This means spatially the two electrons cannot be at the same place which reduces electrostatic repulsion between electrons. Therefore to minimize repulsion we get neighboring spins aligned. This is called ferromagnetic exchange. Excess of aligned spins gives bulk magnetization called Ferromagnetism. Solid materials may lose or gain energy in external magnetic fields and hence are attracted or repelled by magnetic fields and are called paramagnetic and diamagnetic respectively. We say more on this in the text.

We have been talking about electron wavepackets and how they are accelerated by electric field. We now take the opportunity to look at electrons in magnetic field. Consider an electron moving in the 2D plane with magnetic field B applied in the z direction. Then the electron feels the Lorentz force evB perpendicular to its motion and its direction changes. For radius r such that

$$\frac{mv^2}{r} = evB \tag{1.4}$$

the electron executes a circular motion call cyclotron orbit with radius r and velocity $v = \omega r$, such that its angular velocity satisfies

$$\omega = \frac{eB}{m}. \tag{1.5}$$

For a field of $B = 10$ T, we get $\omega \sim 10^{12}$ rad/s.

Now consider a conductor in magnetic field along z direction with current moving along x axis as shown in Fig. 1.17. The electron moving along x axis, feels a Lorentz force along y axis, which leads to development of a hall field E_H (hall voltage V_H) along y direction which balances this force, which gives

$$\frac{V_H}{d} = E_H = evB. \tag{1.6}$$

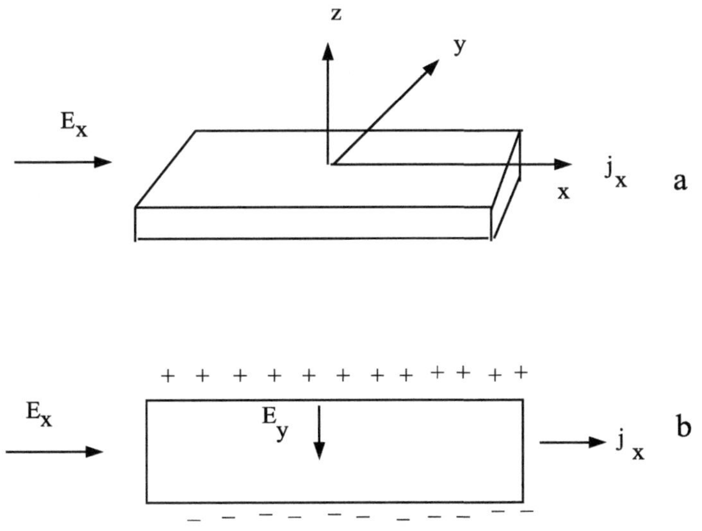

Fig. 1.17 **a** Shows a conductor in magnetic field along z direction with current moving along x axis. **b** Shows the top view of the conductor with Hall field in the y direction

The velocity v is related to current density j_x as $j_x = nev$ where n is the carrier density and $j_x = I/A$, where A is the cross-sectional area of the conductor. Substituting we get

$$R_H = \frac{V_H}{I} = \frac{B}{Ne}, \tag{1.7}$$

where N is total carriers in the conductor. This is called the classical Hall effect. When we put a current carrying conductor in a magnetic field, it develops a Hall voltage. The voltage develops because electron moving along x direction is deflected along y direction. This will make charge accumulate along the top and bottom edge of the conductor as shown in Fig. 1.17b which gives the hall filed E_H and voltage V_H.

On closer inspection, we may find that electrons in the bulk of the 2D plane will be deflected along y direction and they will just execute cyclotron motion. However at edges, we don't have room to execute cyclotron motion. These electrons then press against the edge and develop an Hall field due to electron repulsion. Therefore one may conclude that Hall field is seen only be edge electrons and they give rise to Hall voltage, the bulk doesn't. This phenomenon is revealed in a classical set of experiment called the quantum hall effect we describe in the text.

Finally we mention some experimental techniques that have been developed to study solids. Atoms in solids are spaced few angstrom apart. To study this crystal structure, one can use X-rays which have wavelength in this range. X-rays when scattered from adjacent crystal planes have substantial phase difference between them. Making X-rays scatter of crystal planes and letting them interfere, can shed information on the inter-plane separation. Therefore X-ray crystallography is a major tool to study crystal structures. Instead of using X-rays, we may also use neutrons, which have their de Broglie wavelength same a X-rays, and one can study crystal structures with neutron diffraction. By studying collision of neutrons with phonons, we can learn about phonon dispersion relation and this goes by the name of neutron spectroscopy.

We have talked a bit about electron wavepackets. Let's put down some mathematics describing electron wavepackets that we will use in the rest of the text.

1.2 Wavepackets

Before we discuss the dynamics of localized electron waves, let's get started by talking a bit about electron waves in general. The free electron wavefunction is $\psi = \exp(ikx)$. The momentum is $\frac{\hbar}{i}\frac{\partial}{\partial x}$. This gives the kinetic energy $\epsilon = \frac{p^2}{2m} = \frac{\hbar^2 k^2}{2m}$, which for $\epsilon = \hbar\omega$ gives

$$\omega(k) = \frac{\hbar k^2}{2m}. \tag{1.8}$$

The dispersion is a parabola as shown in Fig. 1.18A.

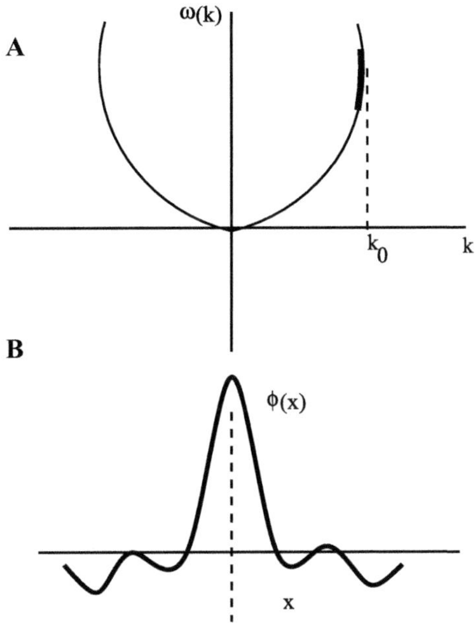

Fig. 1.18 **A** Shows the dispersion $\omega(k)$ versus k for a free electron. **B** Shows a wavepacket centered at k_0

Now, consider a wavepacket centered at k_0, shown the Fig. 1.18A, B. The packet takes the form

$$\phi(x) = \frac{1}{\sqrt{N}} \sum_j \exp(ik_j x), \quad \phi(x,t) = \frac{1}{\sqrt{N}} \sum_j \exp(-i\omega(k_j)t) \exp(ik_j x), \quad (1.9)$$

where $\omega(k_j) = \omega(k_0) + \omega'(k_0)\Delta k_j$ where $\Delta k_j = k_j - k_0$. Denote $v_g = \omega'(k_0) = \frac{\hbar k_0}{m}$, as the group velocity. Then

$$\phi(x,t) = \frac{1}{\sqrt{N}} \exp(i(k_0 x - \omega(k_0)t)) \sum_j \exp(i\Delta k_j (x - v_g t)). \quad (1.10)$$

The function $f(x) = \frac{1}{\sqrt{N}} \sum_j \exp(i\Delta k_j x) = \frac{2}{\sqrt{N}} \sum_j \cos(\Delta k_j x)$, is centered at origin with width $\propto (\Delta k)^{-1}$ as shown in Fig. 1.18B. Then

$$|\phi(x,t)| = |f(x - v_g t)|, \quad (1.11)$$

the free electron wavepacket moves with a group velocity v_g.

Now let's apply an electrical field E in the x direction at $t = 0$. Then the Schröedinger equation is

$$i\hbar\frac{\partial\psi}{\partial t} = \frac{1}{2m}\left(-i\hbar\frac{\partial}{\partial x}\right)^2 + eEx\,\psi. \tag{1.12}$$

The equation is satisfied by time varying wavevectors $\exp(ik(t)x)$, where $k(t) = k - \frac{eEt}{\hbar}$, with energy (dispersion) $\omega(k(t)) = \frac{\hbar(k(t))^2}{2m} = \frac{\hbar(k-\frac{eEt}{\hbar})^2}{2m}$, so that the wavefunction becomes

$$\exp(-i\int_0^t \omega(k(\tau))\,d\tau)\exp(ik(t)x).$$

The initial wavepacket $\phi(x)$ evolves to $\phi(x,t)$, where

$$\phi(x) = \frac{1}{\sqrt{N}}\sum_j \exp(ik_jx),\quad \phi(x,t) = \frac{1}{\sqrt{N}}\sum_j \exp(-i\int_0^t \omega(k_j(\tau))d\tau)\exp(ik_j(t)x). \tag{1.13}$$

The group velocity

$$v_g(t) = \frac{\hbar k(t)}{m} = \frac{\hbar(k-\frac{eEt}{\hbar})}{m};\quad \frac{dv_g(t)}{dt} = -\frac{eE}{m}. \tag{1.14}$$

The electron wavepacket simply accelerates the way we know from classical mechanics. Being more pedagogical, we have

$$\phi(x,t) = \frac{1}{\sqrt{N}}\sum_j \exp\left(-i\int_0^t \omega(k_j(t))\right)\exp(ik_j(t)x) \tag{1.15}$$

$$= \frac{1}{\sqrt{N}}\exp\left(-i\int_0^t \omega(k_0(t))\right)\exp(ik_0(t)x)\sum_j \exp\left(i\Delta k_j\left(x - \int_0^t v_g(\sigma)\right)\right)d\sigma.$$

The wavepacket evolves with instantaneous velocity $v_g(t)$.

The above method can be generalized to arbitrary potential. Consider the Schröedinger equation

$$i\hbar\frac{\partial\psi}{\partial t} = \left(-\frac{\hbar^2}{2m}\frac{\partial}{\partial x^2} - eV(x)\right)\psi. \tag{1.16}$$

We approximate the potential V by piecewise linear potential such that $V(x) = V(x_i) + V'(x_i)\delta x$, where $\delta x = x - x_i$, as shown in Fig. 1.19. We call these regions of linearized potential, cells. We can rewrite the potential in a cell as $V(x) = U(x_i) + V'(x_i)x$.

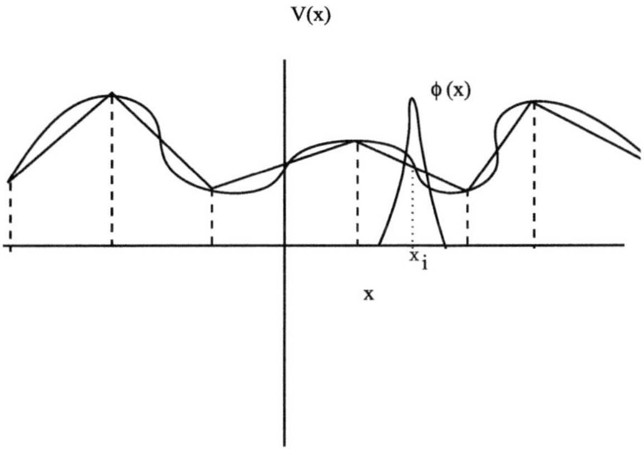

Fig. 1.19 Figure shows linear approximation of potential $V(x)$. The wavepacket $\phi(x)$ is confined to a cell

We assume that the wavepacket has large k_0 such that $\Delta k \sim \sqrt{k_0}$ is large and therefore for the wavepacket, $\Delta x \sim (\Delta k)^{-1}$ is small so that it fits well within one cell. Then in this cell, the Schröedinger equation takes the form

$$i\hbar \frac{\partial \psi}{\partial t} = \left(-\frac{\hbar^2}{2m} \frac{\partial}{\partial x^2} - eV'(x_i)x - eU(x_i) \right) \psi. \tag{1.17}$$

Since the wavepacket is confined to a cell, its evolution would be the same if the potential we have was not only true in the cell but globally true. This is because the wavepacket doesn't know what the potential is outside the cell, it's confined to the cell. Then let's solve the Schröedinger equation with this potential assumed globally true and see how wavepacket evolves.

Then as before the Schröedinger equation is solved by wavevector $\psi = \exp(ik(t)x)$. Let $x(t)$ denote coordinates of the center of wavepacket, then

$$k(t) = k + \frac{e\int_0^t V'(x(\tau))d\tau}{\hbar}, \quad \omega(k(t)) = \frac{\hbar(k + \frac{e\int_0^t V'(x(\tau))d\tau}{\hbar})^2}{2m} - \frac{e\int_0^t U(x(\tau))d\tau}{\hbar}. \tag{1.18}$$

The group velocity

$$v_g(t) = \frac{\hbar(k + \frac{e\int_0^t V'(x(\tau))d\tau}{\hbar})}{m}; \quad \frac{dv_g(t)}{dt} = \frac{eV'(x(t))}{m}. \tag{1.19}$$

This is classical mechanics. Therefore at high energies where k_0 is large and wavepacket is well confined, i.e., over the packet width, the change of potential is small. A linearized potential is a good approximation and evolution in quantum mechanics mimics classical mechanics. We have been talking about free electron wavepackets but everything we have said generalizes directly to electron wavepackets in solids with a dispersion relation $\omega(k)$.

1.3 Temperature and Localized Electron Waves

Why localized electron waves? Just temperature. It is the thermal de Broglie wavelength.

$$\frac{\hbar^2 k \Delta k}{2m} = KT,$$

which gives for $k \sim 10^{10}$ at $T \sim 300\,\text{K}$ is $\Delta k = 10^8$, a packet 100 of nanometers long.

Diatomic molecules are not is ground vibrational states but superposition of eigenstates, vibrating, classical. So are electrons not in electron wavestates but packets and localized electron waves. Just temperature.

1.4 Organization

The book is organized as follows. In Chap. 2, we begin with discussing what holds a solid together. The nature of different kinds of bonding and forces in solids. This is followed by a Chap. 3 on lattice vibrations, called phonons. We show how energy is stored in lattice vibrations, that contribute to specific heat of a solid, and how this energy is exchanged with free electrons in solids. This gives electrons in solid, their Fermi–Dirac distribution, which is developed in this chapter via electron–phonon collisions, which in turn contributes to electronic specific heat of metals. In Chap. 4, we study electron waves in a periodic potential. This chapter develops the theory of energy bands, in which electron waves in solids are organized. In Chap. 5, we apply the band theory of solids to the study of various electronic devices that are all pervasive from consumer electronics and the computer industry to modern communication technology. In Chap. 6, we study an important topic in solid state physics, the subject of superconductivity. Many materials called superconductors exhibit complete loss of electrical resistance when cooled below a characteristic critical temperature. This phenomenon called superconductivity [9, 10] was discovered in mercury by Dutch physicist Onnes in 1911. For decades, a fundamental understanding of this phenomenon eluded the many scientists who were working in the field. Then, in the 1950s and 1960s, a remarkably complete and satisfactory theoretical picture of the classic superconductors emerged in terms of the BCS theory [11]. In this chapter, we develop the BCS theory in terms of localized electron waves. In Chap. 7, we study dynamics of electrons in magnetic fields. This gives rise to an important effect called Hall effect. In Chap. 8, we study another important topic in solid state physics, the subject of magnetism. An everyday example of magnetism is a refrigerator magnet used to hold notes on a refrigerator door. In this chapter, we study the quantum basis for why certain materials (such as iron) form permanent magnets or are attracted to magnets. This phenomenon called ferromagnetism is studied in terms of localized electron waves. We also study other types of magnetism, paramagnetism, diamagnetism, and antiferromagnetism. Finally in Chap. 9 we look at methods by which we can look inside and image periodic structure of atoms in a solid. These include X-ray and neutron diffraction techniques. We then study how spectroscopy using neutrons can be used to study phonons in a solid where we develop the theory of neutron–phonon collisions very much like electron–phonon collisions done earlier.

Problems

1. Electron wave has $k = 10^9$, find its energy and velocity.
2. Find current density j, when $n = 10^{28}/\text{m}^3$ and drift velocity 10^{-3} m.
3. For above find conductivity σ, given relation time $\tau = 10^{-15}$ s.
4. Electron wave with $k_0 = 10^9$ sees an electric field $E = 100$ V/m, for $\tau = 1$ picosecond, find the new k_0.
5. In above find the acceleration of the electron packet when the field is on.

References

1. C. Kittel, *Introduction to Solid State Physics*, 8th edn. (Wiley, 2005)
2. N.W. Ashcroft, D. Mermin, *Solid State Physics* (Harcourt College Publishers, 1976)
3. S. Simon, *Oxford Solid State Basics* (Oxford University Press, 2013)
4. J.R. Hook, H.E. Hall, *Solid State Physics*, 2nd edn. (Wiley, 1995)
5. D.L. Goodstein, *States of Matter* (Dover, 2014)
6. H.M. Rosenberg, *The Solid State* (Oxford University Press, 2000)
7. H. Ibach, H. Luth, *Solid-State Physics* (Springer, 2009)
8. G. Burns, *Solid State Physics* (Academic, 1985)
9. M. Tinkham, *Introduction to Superconductivity*, 2nd edn. (McGraw Hill, 1996)
10. P.G. De Gennes, *Superconductivity of metals and alloys* (Benjamin, Inc, W.A, 1966)
11. J. Bardeen, L. Cooper, J.R. Schriffer, Theory of superconductivity. Phys. Rev. **108**(5), 1175 (1957)

Chapter 2
Bonding in Solids

2.1 Forces and Energies in Solids

What holds a solid together? What is the nature of the cohesive force? In this chapter, we answer some of these questions [1, 2]. When atoms forming a solid are far away, they do not talk to each other, but when brought close, the electrons of one atom feel a positive charge of nuclei of neighboring atom, which lowers its energy. Therefore this becomes a lower energy configuration, compared to atoms far from each other. Hence atoms form solids.

The basic idea is that suppose, we bring the atoms close so that the electron of atom A, enters the electron cloud of electron B. Then the electron of atom A will begin to also see some potential due to ion B, as its charge is not fully screened and this will modify the wavefunction and energies of electron A (Fig. 2.1).

Let us call the potentials of ions V_A and V_B, and the wavefunctions ϕ_A and ϕ_B, each with energy ϵ_0. There is a transition from ϕ_A to ϕ_B given by

$$\langle \phi_B V_B \phi_A \rangle = -t$$

and similarly by symmetry there is the transition from ϕ_B to ϕ_A given by

$$\langle \phi_A V_A \phi_B \rangle = -t$$

$\langle \phi_A V_B \phi_A \rangle$ is small and we neglect. Then in the basis ϕ_A and ϕ_B, we can write the Hamiltonian as

$$H = \begin{bmatrix} \epsilon_0 & -t \\ -t & \epsilon_0 \end{bmatrix}. \tag{2.1}$$

The eigenvalues now are $\epsilon_0 \mp t$, and eigenvectors $\frac{1}{2}(\phi_A \pm \phi_B)$. The two electrons will now occupy the lower energy orbital $\frac{1}{2}(\phi_A + \phi_B)$. The new orbitals are a linear combination of atomic orbital (LCAO) (Fig. 2.2).

N. Khaneja, *Electron Waves in Solids*, Springer Series in Solid-State Sciences 203,
https://doi.org/10.1007/978-3-031-67260-6_2

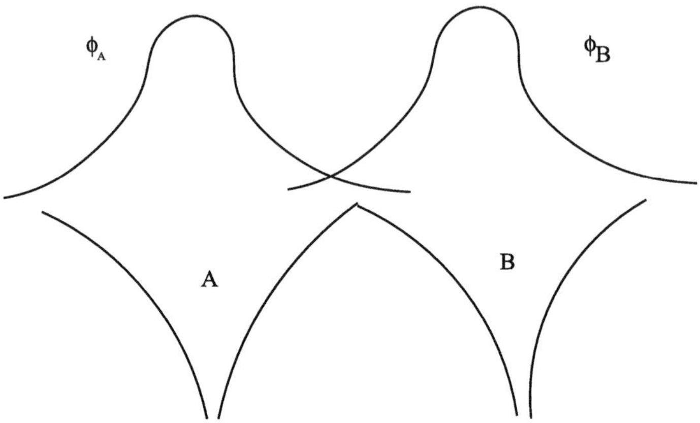

Fig. 2.1 Figure shows two potential wells A and B and their corresponding orbitals with overlap between them

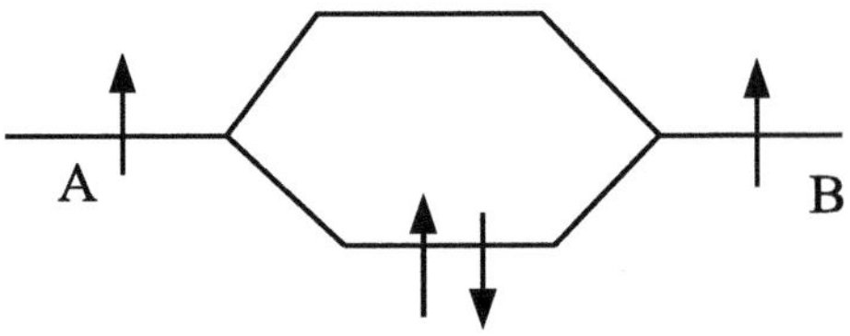

Fig. 2.2 Figure shows two atomic orbitals ϕ_A and ϕ_B and their linear combinations with energy $\epsilon_0 - t$ and $\epsilon_0 + t$

Thus, we have seen how atomic orbitals are modified when we bring atoms closer. We form molecular orbitals and a bond. When energies of ϕ_A and ϕ_B (written as ϵ_0) are comparable, we form a covalent bond. When they are very different as in common salt NaCl, we form a so-called ionic bond.

In this case, the Hamiltonian in (2.2) gets modified with energies $\epsilon_A = \epsilon_0 + \Delta$ and $\epsilon_B = \epsilon_0 - \Delta$

$$H = \begin{bmatrix} \epsilon_A & -t \\ -t & \epsilon_B \end{bmatrix}. \tag{2.2}$$

The eigenvalues are $\epsilon_1 = \epsilon_0 - \sqrt{t^2 + \Delta^2}$ and $\epsilon_2 = \epsilon_0 + \sqrt{t^2 + \Delta^2}$, with eigenvectors $e_1 = \cos\theta\phi_A + \sin\theta\phi_B$ and $e_2 - \sin\theta\phi_A + \cos\theta\phi_A$, with $\sin\theta = \frac{\Delta}{\sqrt{t^2+\Delta^2}}$. Figure 2.3 shows two atomic orbitals ϕ_A and ϕ_B, with energies $\epsilon_A = \epsilon_0 + \Delta$ and $\epsilon_B = \epsilon_0 - \Delta$, and their linear combinations with energy $\epsilon_0 - \sqrt{t^2 + \Delta^2}$ and $\epsilon_0 + \sqrt{t^2 + \Delta^2}$. When $\Delta \gg t$, we have $\sin\theta \sim 1$ and $e_1 \sim \phi_B$, i.e., the lower energy molecular orbital is simply ϕ_B. Then we say atom A donates an electron and it goes to B.

Until now, our discussion has been limited to covalent and ionic bonds between two atoms, and how it reduces energy. But that would say, we should form a gas of diatomic molecules. Why do we instead end up forming a solid?

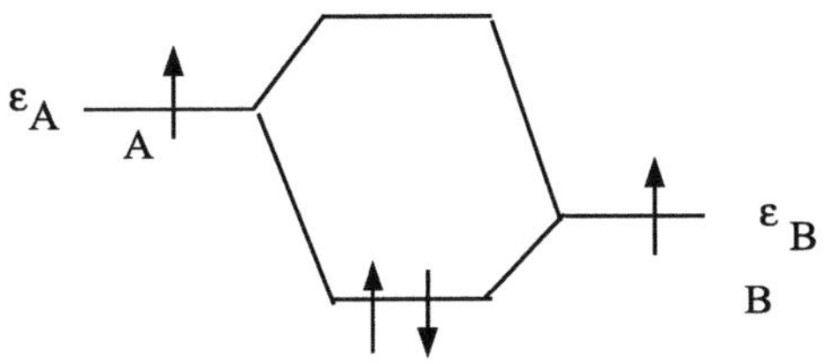

Fig. 2.3 Figure shows two atomic orbitals ϕ_A and ϕ_B, with energies $\epsilon_A = \epsilon_0 + \Delta$ and $\epsilon_B = \epsilon_0 - \Delta$, and their linear combinations, with energy $\epsilon_1 = \epsilon_0 - \sqrt{t^2 + \Delta^2}$ and $\epsilon_2 = \epsilon_0 + \sqrt{t^2 + \Delta^2}$

2.2 Ionic and Covalent Solids

Ionic bonds, between atoms $A - B$, move charge from A to B and creates A^+B^-. Now consider a solid configuration, where A^+ is surrounded by B^-, and has A^+ as its second ngbs, and B^- the third ngbs, as shown in Fig. 2.4.

In this regular arrangement, A ions feel attractive potential due to ngbs B ions, and feel repulsive potential due to second ngbs A ions, that are farther and then attractive potential due to thirds ngbs B ions and so on. In nutshell, we have an alternating series of the kind $V(R) = -c(1/R - 1/2R + 1/3R + \dots)$ with c appropriate constant, which in the end gives an attractive potential. We can write it as $V(R) = -\frac{\alpha}{R}$, where constant α goes by the name of *Madelung* constant. The resulting attractive potential lowers the total energy and explains why we have ionic solids. Prime examples are sodium chloride NaCl, with $A = Na$ and $B = Cl$, or lithium fluoride LiF, with $A = Li$ and $B = F$. Figure 2.5 shows arrangement of $A - B$ ions, with A as sodium and B as chlorine. B ions are arranged on vertices, and center of cube faces (so-called fcc lattice), and so is A, with the two fcc lattices displaced with respect to each other.

How about covalent solids? In this case, there is no charge transfer. Why do we have covalent bonded diatomic molecules packed in a solid? The answer is we have

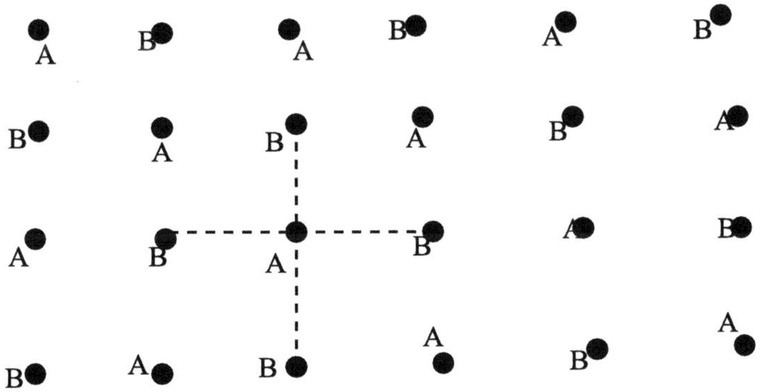

Fig. 2.4 Figure shows regular arrangement of $A - B$ ions with B as immediate ngbs of A

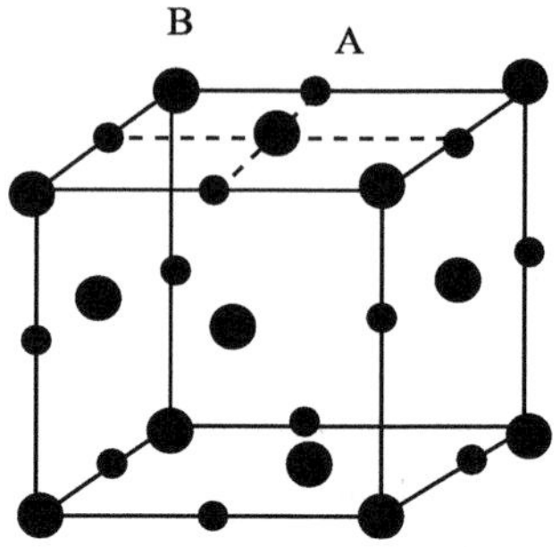

Fig. 2.5 Figure shows regular arrangement of $A - B$ ions, with A as sodium and B as chlorine. B ions are arranged on vertices, and center of cube faces, and so is A, with the two lattices displaced with respect to each other

covalent bonds, with all ngbs and they have bonds with all their ngbs, and everything gets connected. For example, consider silicon with outer electrons in configuration $3s^2 3p^2$. The one s, and three p orbitals, hybridize to form four sp^3 orbitals. Silicon has four ngbs, and we use one sp^3 orbital for each ngb and form a bond as shown in Fig. 2.6. More precisely, Si atoms are arranged on a fcc lattice, and a displaced one with $(0, 0, 0)$ displaced to $(\frac{1}{4}, \frac{1}{4}, \frac{1}{4})$. Then $(\frac{1}{4}, \frac{1}{4}, \frac{1}{4})$ bonds to $(0, 0, 0)$, $(0, \frac{1}{2}, \frac{1}{2})$ and $(\frac{1}{2}, \frac{1}{2}, 0)$ and $(\frac{1}{2}, 0, \frac{1}{2})$ as shown in Fig. 2.7.

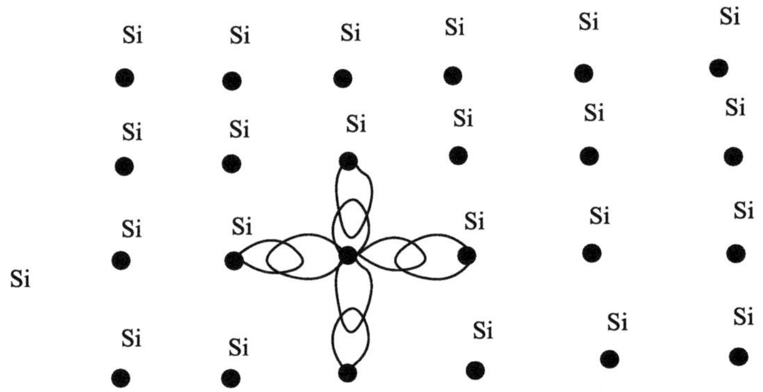

Fig. 2.6 Figure shows the regular arrangement of Si atoms forming covalent bands with its ngbs

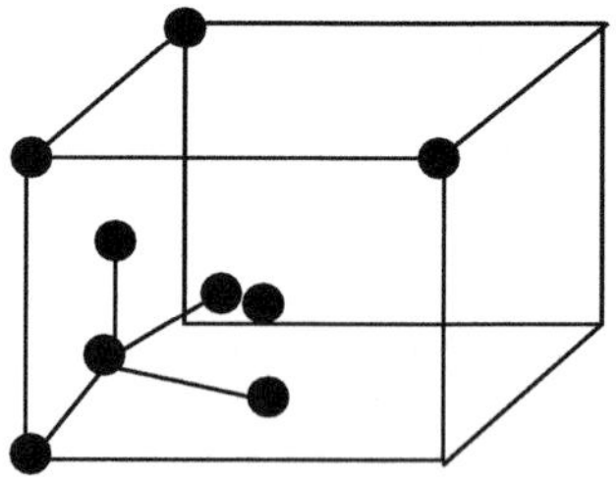

Fig. 2.7 Figure shows how Si atoms are arranged on an fcc lattice and a displaced one with $(0, 0, 0)$ displaced to $(\frac{1}{4}, \frac{1}{4}, \frac{1}{4})$. Then $(\frac{1}{4}, \frac{1}{4}, \frac{1}{4})$ bonds to $(0, 0, 0)$, $(0, \frac{1}{2}, \frac{1}{2})$ and $(\frac{1}{2}, \frac{1}{2}, 0)$ and $(\frac{1}{2}, 0, \frac{1}{2})$

We talked about Si, another example is Gallium Arsenide, GaAs, with Gallium ad Arsenic on $(\frac{1}{4}, \frac{1}{4}, \frac{1}{4})$ and $(0, 0, 0)$ fcc lattices. Both Gallium ad Arsenic are sp^3 hybridized and form bonds. Energy of these two sp^3 is not same, with Gallium more electropositive hence at higher energy, therefore the bond has some ionic character as discussed in ionic bonding.

2.3 Metallic Bonding

In our discussion of ionic and covalent solids, we first formed a bond and then understood how diatomic molecules are arranged as a solid. In metals, instead of looking at a molecular bond, formed from two atoms, we look at a molecular bond among n atoms. Consider a periodic array of atoms, say first in one dimension, also called a 1 d chain (Fig. 2.8).

Electron on 1 is scattered by potential on 2, on 2 by 3, and so on. The resulting Hamiltonian has the form

$$H = \begin{bmatrix} \epsilon_0 & -t & 0 & \dots\dots & & -t \\ -t & \epsilon_0 & -t & 0 & \ddots & \vdots \\ 0 & -t & \epsilon_0 & -t & 0 & \vdots \\ \vdots & \ddots & \ddots & \ddots & \ddots & \vdots \\ 0 & \dots\dots & & -t & \epsilon_0 & -t \\ -t & 0 & \dots\dots & & -t & \epsilon_0 \end{bmatrix}, \tag{2.3}$$

where we have introduced a transition between 1 and n to close the chain. This is an approximation, where we have approximated a very large (infinite) size H as a limit of the circulant matrix. H above is a circulant matrix of the form

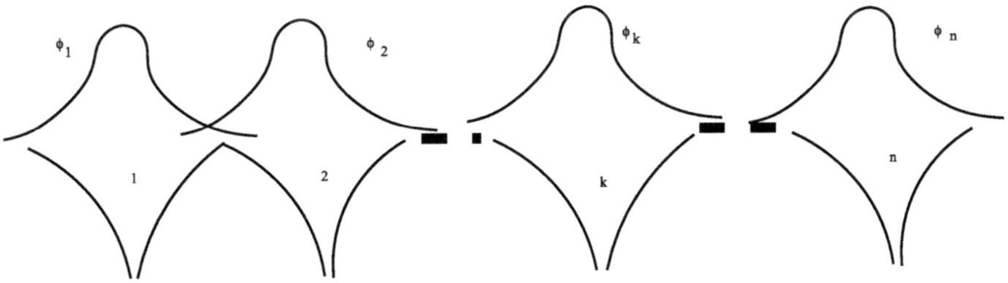

Fig. 2.8 Figure shows atomic orbitals ϕ_1 to ϕ_n and their potential wells

$$\begin{bmatrix} r_0 & r_1 & r_2 & \ldots\ldots & r_n \\ r_n & r_0 & r_1 & r_2 & \ddots & \vdots \\ r_{n-1} & r_n & r_0 & r_1 & r_2 & \vdots \\ \vdots & \ddots & \ddots & \ddots & \ddots & \vdots \\ r_2 & \ldots\ldots & r_n & r_0 & r_1 \\ r_1 & r_2 & \ldots\ldots & r_n & r_0 \end{bmatrix}. \tag{2.4}$$

A circulant matrix is always diagonalized by a DFT matrix whose jth column is $\frac{1}{\sqrt{n}}\begin{pmatrix} 1 \\ \omega^{j-1} \\ \omega^{2(j-1)} \\ \ldots \\ \omega^{(n-1)(j-1)} \end{pmatrix}$ with eigenvalue $\sum_k r_k(\omega^{j-1})^{k-1}$ where ω in nth root of unity $\omega = \exp(i\frac{2\pi}{n})$. Then our eigenvalues are

$$\epsilon_0 - 2t\cos\frac{2\pi j}{n}, \quad 0 < j < n-1,$$

which can be written as

$$\epsilon(k) = \epsilon_0 - 2t\cos ka, \quad \frac{\pi}{a} < k < \frac{\pi}{a}. \tag{2.5}$$

This is called a dispersion relation. The eigen vectors are of the form

$$\Psi(x) = \sum_l \exp(ikla)\phi(x - la). \tag{2.6}$$

Thus starting from isolated atomic orbitals, with one energy ϵ_0, we branch into many energies as we decrease the spacing between atoms. The energy gets broadened into a band. This is shown in figure (Fig. 2.9).

Now coming back to 1 d chain of hydrogen-like atoms, we found that given N atoms, we get N wavevectors ranging from $\frac{-\pi}{a}$ to $\frac{\pi}{a}$, we have N orbitals with

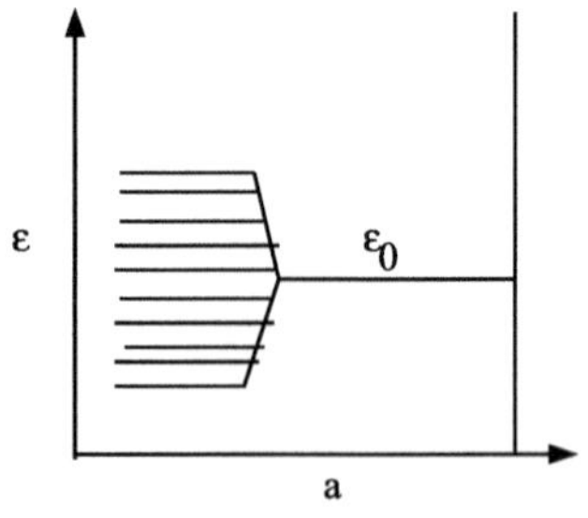

Fig. 2.9 Figure shows how atomic orbital energy ϵ_0 gets branched into many energies as we decrease the separation a between orbitals

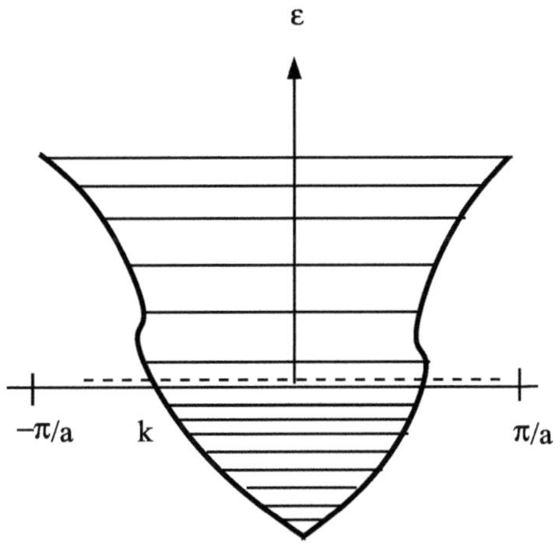

Fig. 2.10 Figure shows a plot of dispersion relation in (2.5)

separation of $\frac{2\pi}{N}$ between them. These are shown in Fig. 2.10. By the Pauli exclusion principle, each orbital can carry only 2 electrons with opposite spins. Since we have N electrons, if we fill the energy levels starting from minimum energy, we get only $\frac{N}{2}$ filled orbitals as each can hold two electrons. The remaining $\frac{N}{2}$ orbitals are empty. Thus for positive t, all orbitals have smaller energy than ϵ_0, as $2t\cos ka > 0$ for $\frac{\pi}{2a} < k < \frac{\pi}{2a}$. Therefore we reduce energy by putting metal atoms in a regular arrangement and this is a metallic bond that holds metal together.

The tight-binding approximation has been described using a 1D chain. In real solids, we have a periodic arrangement of ions in three dimensions. Then we have to solve tight-binding approximation for three dimensions. metallic bonding is found in metals like sodium, magnesium, aluminum, etc.

2.4 Van der Waals Solids

Noble gases like neon, argon, and xenon at low temperatures form solid. What kind of forces hold them together as their electronic configuration is filled with shells. The forces that are active in these solids is the Van der Waals force of attraction.

We begin by recapitulating the Van der Waals interaction. Consider two identical inert gas atoms (Helium atoms). We model them by identical linear harmonic oscillators 1 and 2, separated by distance R. Each oscillator bears charges $\pm e$, with separation x_1, x_2 as in Fig. 2.11. The oscillators oscillate along x axis, with momenta p_1, p_2. The force constant is C. Then the Hamiltonian of the unperturbed system is

$$H_0 = \frac{p_1^2}{2m} + \frac{p_2^2}{2m} + \frac{Cx_1^2}{2} + \frac{Cx_2^2}{2}. \tag{2.7}$$

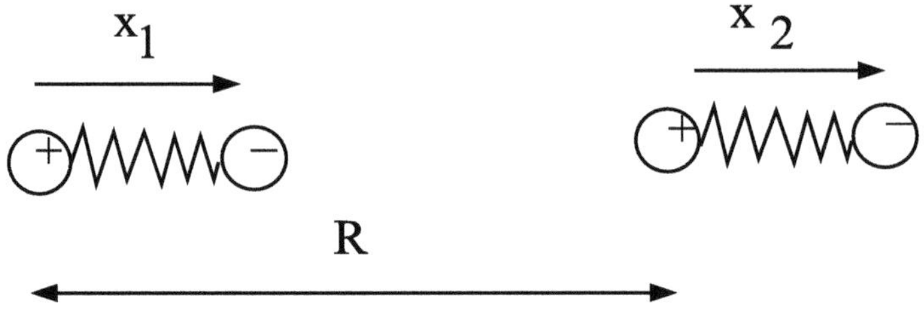

Fig. 2.11 Figure depicts how electrons in two Helium atoms are modeled as spring mass system

Each uncoupled oscillator is assumed to have frequency $C = m\omega_0^2$. The resulting dipoles interact with a potential

$$H_1 = -\frac{1}{2\pi\epsilon_0}\frac{e^2 x_1 x_2}{R^3}. \tag{2.8}$$

Writing normal modes for this problem with $x_s = \frac{x_1+x_2}{\sqrt{2}}$, $x_a = \frac{x_1-x_2}{\sqrt{2}}$ and $p_s = \frac{p_1+p_2}{\sqrt{2}}$ and $p_s = \frac{x_1-x_2}{\sqrt{2}}$, we get

$$H_0 + H_1 = \frac{p_s^2}{2m} + \left(C - \frac{e^2}{2\pi\epsilon_0 R^3}\right)\frac{x_s^2}{2} + \frac{p_a^2}{2m} + \left(C + \frac{e^2}{2\pi\epsilon_0 R^3}\right)\frac{x_a^2}{2} \tag{2.9}$$

with frequencies

$$\omega = m^{-\frac{1}{2}}\left(C \pm \frac{e^2}{2\pi\epsilon_0 R^3}\right)^{\frac{1}{2}} = \omega_0\left(1 \pm \frac{e^2}{4\pi\epsilon_0 R^3 C} - \frac{1}{2}\left(\frac{e^2}{4\pi\epsilon_0 R^3 C}\right)^2 + \ldots\right). \tag{2.10}$$

Thus the zero-point energy of the system changes by

$$\Delta U = \frac{1}{2}\hbar(\Delta\omega_a + \Delta\omega_s) = -\frac{\hbar\omega_0}{2}\left(\frac{e^2}{4\pi\epsilon_0 R^3 C}\right)^2 = -\frac{A}{R^6}. \tag{2.11}$$

This is the Van der Waals attractive potential.

To put some numbers, Fig. 2.12 shows a helium atom with an electron cloud displaced. This displacement Δx of electron cloud increases Coulomb energy and gives a restoring force with spring constant C. We estimate C by finding how much is increase in coulomb energy and equating it to $\frac{1}{2}C\Delta x^2$. This gives $C \sim \frac{e^2}{a_0^3 4\pi\epsilon_0}$, where e and a_0 are the charge and radius of Helium cloud. We get $C \sim 10^4$ J/m^2 and $A \sim 10^{-80}$ J m^6.

The Van der Waals attractive force goes as A/R^6. When two inert gas atoms are brought together at very small distances they develop a repulsive force, as their electron wavefunction overlap. By Pauli exclusion principle, this is not allowed and

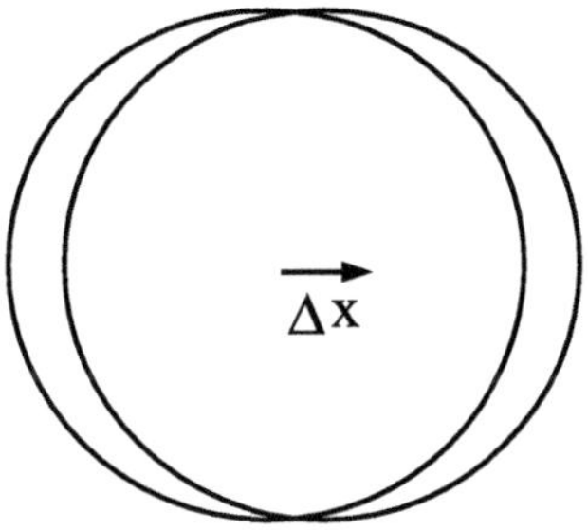

Fig. 2.12 Figure depicts how displacement Δx of electron cloud increases energy and gives a restoring force with spring constant C

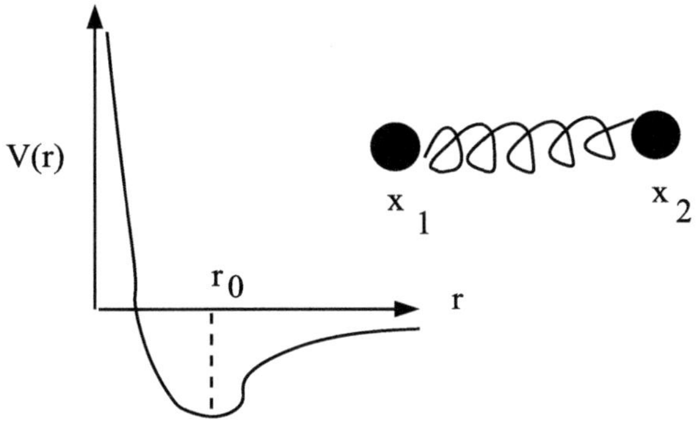

Fig. 2.13 Figure depicts Lenard Jones potential between two inert gas atoms, which is attractive at large distances due to Van der Waals forces and becomes repulsive at small distances

results in the promotion of electrons to higher energy orbitals which leads to an increase of energy and hence a repulsive force. For inert gases, a force law B/r^{12} models this well and hence we have a total potential (also called Lenard Jones potential) of the form

$$V(r) = \frac{B}{r^{12}} - \frac{A}{r^6}. \tag{2.12}$$

This is plotted in Fig. 2.13. It achieves its minimum at a typical distance $r_0 \sim 2 - 3$ A°. When inert gas atoms like neon, argon, and xenon are cooled to low temperatures with densities with an inter-atomic distance of r_0, the inert gas atoms condense to form a solid. These are solid held together with a cohesive force with origin as Van der Waals attraction.

Problems

1. For $\epsilon_0 = -4$ eV and $t = 2$ eV, find the energy of bonding and antibonding orbital for a covalent bond.
2. For $\epsilon_A = -4$ eV, $\epsilon_B = -4.5$ eV and $t = 2$ eV, find energy of bonding and antibonding orbital for a ionic bond $A - B$.

3. How wide (in energy units) is a band for $t = 2$ eV.
4. In Fig. 2.5, estimate Madulung constant α.
5. Estimate Van der Waal potential A from spring constant $C = 10^4\ \mathrm{J/m^2}$.

References

1. J.N. Murrell, S.F.A. Kettle, J.M. Tedder, *The Chemical Bond* (Wiley, 1985)
2. L. Pauling, *The Nature of the Chemical Bond and the Structure of Molecules and Crystals* (Cornell University Press, 1940)

Chapter 3
Phonons in Solids

3.1 Phonons

Consider two atoms in a solid, with their inter-nuclear potential plotted in Fig. 3.1B. The potential is minimum at the equilibrium distance r_0. When atoms are displaced from their locations by amount x_1, x_2, we get a change in potential energy

$$\Delta V = \frac{1}{2}V''(r_0)(x_1 - x_2)^2 = \frac{1}{2}k(x_1 - x_2)^2. \tag{3.1}$$

Then equation of motion for atoms in Fig. 3.1A in the 1D configuration is

$$\frac{d^2}{dt^2}\begin{bmatrix} x_{-n} \\ \vdots \\ x_0 \\ \vdots \\ x_n \end{bmatrix} = \beta^2 \underbrace{\begin{bmatrix} -2 & 1 & 0 & \dots & 0 \\ 1 & -2 & 1 & 0 & \vdots \\ \vdots & \ddots & \ddots & \ddots & \vdots \\ \vdots & \dots & 1 & -2 & 1 \\ 0 & 0 & 0 & 1 & -2 \end{bmatrix}}_{A} \begin{bmatrix} x_{-n} \\ \vdots \\ x_0 \\ \vdots \\ x_n \end{bmatrix} \tag{3.2}$$

with $\beta^2 = \frac{k}{m}$. Instead of matrix A, consider

$$\frac{d^2}{dt^2}\begin{bmatrix} x_{-n} \\ \vdots \\ x_0 \\ \vdots \\ x_n \end{bmatrix} = \beta^2 \underbrace{\begin{bmatrix} -2 & 1 & 0 & \dots & 1 \\ 1 & -2 & 1 & 0 & \vdots \\ \vdots & \ddots & \ddots & \ddots & \vdots \\ \vdots & \dots & 1 & -2 & 1 \\ 1 & 0 & 0 & 1 & -2 \end{bmatrix}}_{B} \begin{bmatrix} x_{-n} \\ \vdots \\ x_0 \\ \vdots \\ x_n \end{bmatrix}, \tag{3.3}$$

N. Khaneja, *Electron Waves in Solids*, Springer Series in Solid-State Sciences 203,
https://doi.org/10.1007/978-3-031-67260-6_3

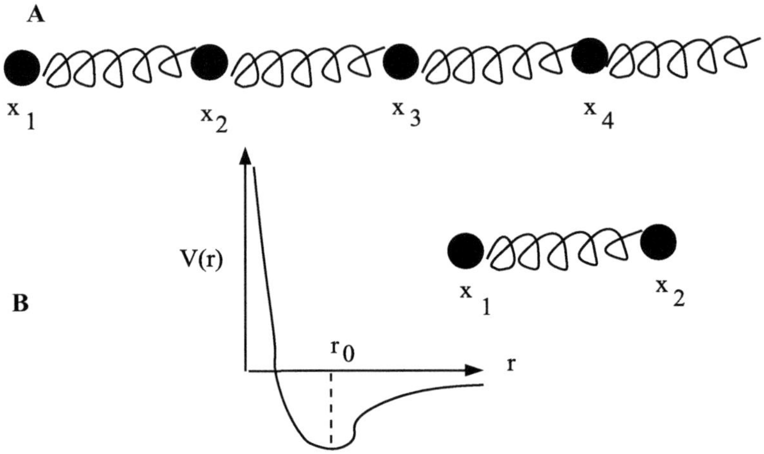

Fig. 3.1 **A** Shows atomic chain, as masses coupled with springs. **B** Shows two such masses, coupled with springs

which corresponds to a closed system.

Both matrix A and B, in limit of $n \to \infty$ are same. We study the solution of B keeping the limit in mind. Observe B is a circulant matrix and for $N = 2n + 1$ and $k = \frac{2\pi m}{Na}$, with $-n \leq m \leq n$, eigenfunction of B is of the form

$$e = \frac{1}{\sqrt{N}} \begin{bmatrix} \exp(-inka) \\ \vdots \\ \exp(-ika) \\ 1 \\ \exp(ika) \\ \vdots \\ \exp(inka) \end{bmatrix} \tag{3.4}$$

with eigenvalues $4\sin^2 \frac{ka}{2}$. The above solution is simply, a wave solution $\exp(ikx)$, with wavevector k, and then the solution to (3.3) is simply a traveling wave $\exp(i(kx - \omega t))$, with

$$|\omega| = 2\beta \left|\sin \frac{ka}{2}\right| \sim \upsilon|k|, \tag{3.5}$$

where $\upsilon = a\beta$, the velocity of sound. This dispersion relation is plotted in Fig. 3.2. In the remaining part of this section, we generalize all this to three dimensions, and study phonons for different lattice structures [1, 2].

Now consider atoms in three dimensions, arranged in a simple cubic structure, which is to say on vertices of a cube of size a as in Fig. 3.3.

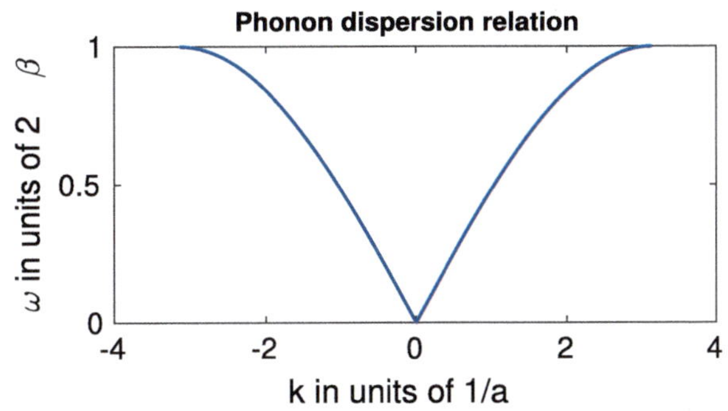

Fig. 3.2 Figure shows phonon dispersion curve

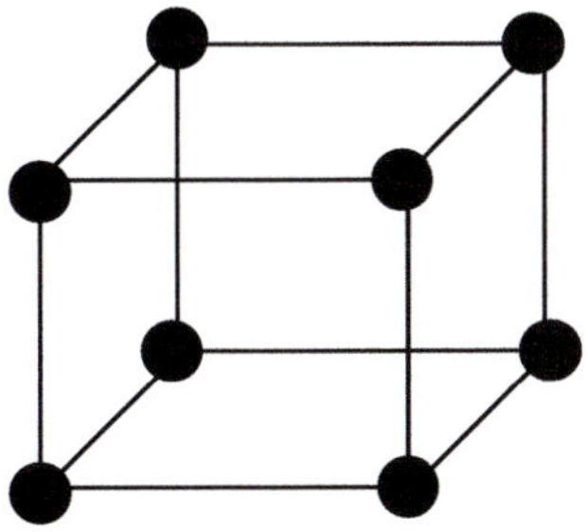

Fig. 3.3 Figure shows a simple cubic lattice, with atoms at the corner points

Consider x displacement, which now we index by three index $u_{k,l,m}$ ($v_{k,l,m}$ and $w_{k,l,m}$ for y and z displacement), with $-n \leq k, l, m \leq n$. Then

$$\begin{aligned} \ddot{u}_{k,l,m} = \beta^2 \Bigg(& u_{k+1,l,m} + u_{k-1,l,m} - 2u_{k,l,m} \\ & + \frac{u_{k-1,l+1,m} + u_{k-1,l+1,m} + u_{k-1,l-1,m} + u_{k+1,l-1,m}}{2} - 2u_{k,l,m} \\ & + \frac{v_{k+1,l+1,m} - v_{k-1,l+1,m} - v_{k+1,l-1,m} + v_{k-1,l-1,m}}{2} \\ & + \frac{u_{k-1,l,m+1} + u_{k+1,l,m+1} + u_{k-1,l,m-1} + u_{k+1,l,m-1}}{2} - 2u_{k,l,m} \\ & + \frac{w_{k+1,l,m+1} - w_{k-1,l,m+1} - w_{k+1,l,m-1} + w_{k-1,l,m-1}}{2} \Bigg). \end{aligned} \tag{3.6}$$

Let us sketch the derivation of (3.6). Let r be the distance between points k, l, m and k', l', m' (nominal value r_0). The x-displacement $u_{k,l,m}$ and $u_{k',l',m'}$ with $\Delta u = u_{k',l',m'} - u_{k,l,m}$ produces $\Delta r = \cos\theta \Delta u$, where θ is angle $\mathbf{r}$ makes with x axis. From (3.1), we get $\Delta V = \frac{1}{2} V''(r_0) \Delta r^2$, which gives $k = V''(r_0)$ as spring constant, which gives a restoring force $F_r = -k\Delta r$ along $\mathbf{r}$ direction which is resolved along x direction as $F_x = \cos\theta F_r$ to give

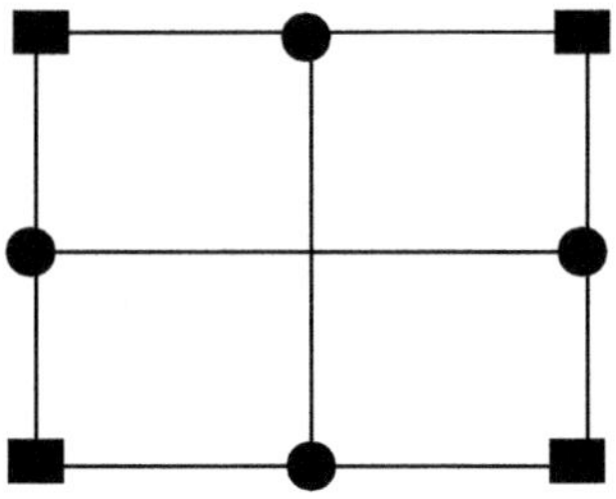

Fig. 3.4 Figure shows ngbs, as squares and circles

$$\ddot{u}_{k,l,m} = \sum V''(r_0)\cos^2\theta \Delta u = \sum_{k',l',m'} V''(r_0)\cos^2\theta (u_{k',l',m'} - u_{k,l,m}). \tag{3.7}$$

Now we sum the RHS in the above equation for all k', l', m'. We in practice only include the nearest or next nearest nghbs.

For example in Fig. 3.4 if $u_{k,l,m}$ is in center, then the square nghs give $\cos^2\theta = \frac{1}{2}$, the horizontal circles gives $\cos^2\theta = 1$ and $\cos^2\theta = 0$ for vertical circles. Then on substitution in (3.6) we get

$$\begin{aligned}
\ddot{u}_{k,l,m} = \beta^2(&(u_{k+1,l,m} + u_{k-1,l,m} - 2u_{k,l,m}) \\
&+ \left(\frac{u_{k+1,l+1,m} + u_{k-1,l+1,m}}{2} + \frac{u_{k-1,l-1,m} + u_{k+1,l-1,m}}{2} - 2u_{k,l,m} \right) \\
&+ \left(\frac{u_{k-1,l,m+1} + u_{k+1,l,m+1}}{2} + \frac{u_{k-1,l,m-1} + u_{k+1,l-1,m-1}}{2} - 2u_{k,l,m} \right) \\
&+ v \; terms + \; w \; terms \;).
\end{aligned}$$

Now writing $\frac{u_{k+1,l+1,m}+u_{k-1,l+1,m}}{2} = \frac{u_{k+1,l+1,m}+u_{k-1,l+1,m}-2u_{k,l+1,m}}{2} + u_{k,l+1,m}$, gives (3.6).

In (3.6), dividing and multiplying by a^2 and taking limit of small a, the equation takes the form ($\upsilon = \beta a$)

$$\frac{\partial^2 u}{\partial t^2} = \upsilon^2 \left(3\frac{\partial^2 u}{\partial x^2} + \frac{\partial^2 u}{\partial y^2} + \frac{\partial^2 u}{\partial z^2} + 2\frac{\partial^2 v}{\partial x \partial y} + 2\frac{\partial^2 w}{\partial x \partial z} \right). \tag{3.8}$$

Similarly, we have

$$\frac{\partial^2 v}{\partial t^2} = \upsilon^2 \left(3\frac{\partial^2 v}{\partial y^2} + \frac{\partial^2 v}{\partial x^2} + \frac{\partial^2 v}{\partial z^2} + 2\frac{\partial^2 u}{\partial x \partial y} + 2\frac{\partial^2 w}{\partial x \partial z} \right). \tag{3.9}$$

$$\frac{\partial^2 w}{\partial t^2} = \upsilon^2 \left(3\frac{\partial^2 w}{\partial z^2} + \frac{\partial^2 w}{\partial x^2} + \frac{\partial^2 w}{\partial y^2} + 2\frac{\partial^2 u}{\partial x \partial z} + 2\frac{\partial^2 v}{\partial y \partial z} \right). \tag{3.10}$$

Then a direct substitution of the wave solution

$$u = u_0 \exp(ik_x x) \exp(-i\omega t) \tag{3.11}$$

with $-\frac{\pi}{a} \leq k_x \leq \frac{\pi}{a}$ and

$$\omega = \sqrt{3}\upsilon |k_x| \tag{3.12}$$

constitutes a longitudinal phonon traveling in $[1, 0, 0]$ direction. As in one 1D, we have a real traveling wave solution given by $\cos(k_x x - \omega t)$, traveling in x direction. Similarly

$$u = u_0 \exp(ik_y y) \exp(-i\omega t) \tag{3.13}$$

with $-\frac{\pi}{a} \leq k_y \leq \frac{\pi}{a}$ and

$$\omega = \upsilon |k_y|, \tag{3.14}$$

constitutes a transverse phonon traveling in $[0, 1, 0]$ direction.

How does a phonon in $\mathbf{k}' = (k_x, k_y, k_z)$ direction look? Let

$$u = A \exp(ik_x x + k_y y + k_z z) \exp(-i\omega t); \quad v = B \exp(ik_x x + k_y y + k_z z) \exp(-i\omega t);$$
$$w = C \exp(ik_x x + k_y y + k_z z) \exp(-i\omega t), \tag{3.15}$$

with $-\frac{\pi}{a} \leq k_x, k_y, k_z \leq \frac{\pi}{a}$. Then from (3.8), (3.9), and (3.10), we have

$$\omega^2 \begin{bmatrix} A \\ B \\ C \end{bmatrix} = \upsilon^2 \left(2\mathbf{k}\mathbf{k}' + |k|^2\right) \begin{bmatrix} A \\ B \\ C \end{bmatrix}. \tag{3.16}$$

For nontrivial solution $\begin{bmatrix} A \\ B \\ C \end{bmatrix} = \mathbf{k}$ and $\omega = \sqrt{3}\upsilon|k|$, is a longitudinal phonon. $\begin{bmatrix} A \\ B \\ C \end{bmatrix} = \mathbf{k}^{\perp}$ is a transverse phonon with $\omega = \upsilon|k|$.

The dispersion curves for longitudinal and transverse phonons differ as shown in Fig. 3.5.

We calculated phonons for a cubic lattice, we now consider an fcc lattice as shown in Fig. 3.6. The lattice is spanned by basis vectors $a_1 = \frac{\hat{x}+\hat{y}}{2}$ and $a_2 = \frac{\hat{y}+\hat{z}}{2}$ and $a_3 = \frac{\hat{x}+\hat{z}}{2}$. The index k, l, m denotes the point $a(ka_1 + la_2 + ma_3)$. A point has 12 ngbhs, 4 in each plane. They all look like squares as in Fig. 3.4. The displacement u, v, w at lattice point k, l, m along x, y, z direction then satisfies the equation ($\upsilon = \frac{\beta a}{2}$)

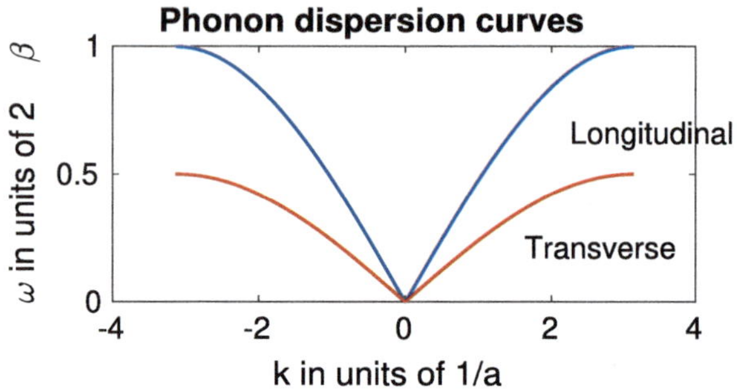

Fig. 3.5 Figure shows dispersion curve for longitudinal and transverse phonon

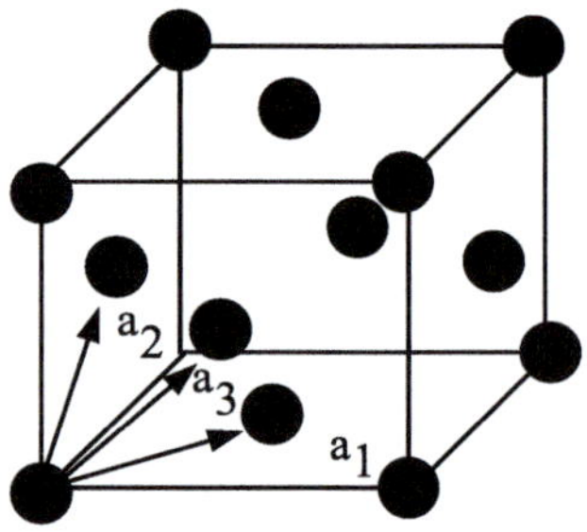

Fig. 3.6 Figure shows a fcc lattice, with atoms at corners and center of faces of a cube

$$\frac{\partial^2 u}{\partial t^2} = \upsilon^2\left(2\frac{\partial^2 u}{\partial x^2} + \frac{\partial^2 u}{\partial y^2} + \frac{\partial^2 u}{\partial z^2} + 2\frac{\partial^2 v}{\partial x \partial y} + 2\frac{\partial^2 w}{\partial x \partial z}\right). \tag{3.17}$$

Similarly, we have

$$\frac{\partial^2 v}{\partial t^2} = \upsilon^2\left(2\frac{\partial^2 v}{\partial y^2} + \frac{\partial^2 v}{\partial x^2} + \frac{\partial^2 v}{\partial z^2} + 2\frac{\partial^2 u}{\partial x \partial y} + 2\frac{\partial^2 w}{\partial x \partial z}\right). \tag{3.18}$$

$$\frac{\partial^2 w}{\partial t^2} = \upsilon^2\left(2\frac{\partial^2 w}{\partial z^2} + \frac{\partial^2 w}{\partial x^2} + \frac{\partial^2 w}{\partial y^2} + 2\frac{\partial^2 u}{\partial x \partial z} + 2\frac{\partial^2 v}{\partial y \partial z}\right). \tag{3.19}$$

How does a phonon in $\mathbf{k} = (K_x, K_y, K_z)$ direction look? Let

$$u = A\exp(iK_xX + K_yY + K_zZ)\exp(-i\omega t);\quad v = B\exp(iK_xX + K_yY + K_zZ)\exp(-i\omega t);$$
$$w = C\exp(iK_xX + K_yY + K_zZ)\exp(-i\omega t). \tag{3.20}$$

This is a solution in real space. The real space coordinates $\mathbf{r} = (X, Y, Z)$ has primitive coordinates (x, y, z) where the point $\mathbf{r} = xa_1 + ya_2 + za_3$. The $\mathbf{k} = (K_x, K_y, K_z)$ in solution

$$\exp(i\mathbf{k}\cdot\mathbf{r})\exp(-i\omega t) \tag{3.21}$$

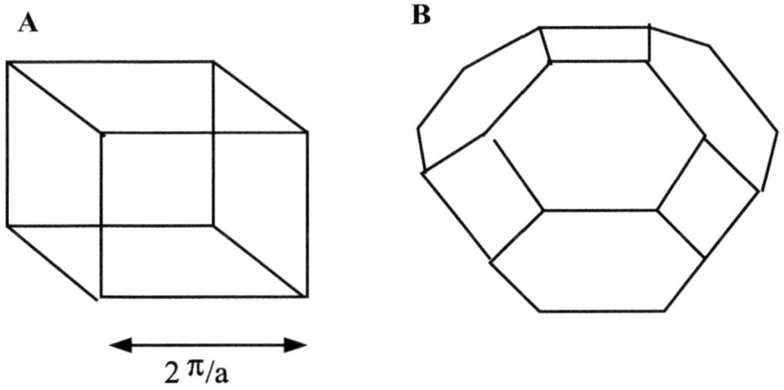

Fig. 3.7 Figure shows the region where the k vector lies (so called the Brillouin zone), for cubic (**A**) and fcc lattice (**B**)

has corresponding primitive wavevectors (k_x, k_y, k_z) where

$$\mathbf{k} = k_x b_1 + k_y b_2 + k_z b_3 \tag{3.22}$$

with b_i reciprocal lattice vectors given by $b_1 = \frac{a_2 \times a_3}{a_1 \cdot (a_2 \times a_3)}$, $b_2 = \frac{a_3 \times a_1}{a_2 \cdot (a_3 \times a_1)}$ and $b_3 = \frac{a_1 \times a_2}{a_3 \cdot (a_1 \times a_2)}$, which gives $b_1 = \hat{x} - \hat{y} + \hat{z}$, $b_2 = \hat{x} + \hat{y} - \hat{z}$ and $b_3 = -\hat{x} + \hat{y} + \hat{z}$ with $-\frac{\pi}{a} \leq k_x, k_y, k_z \leq \frac{\pi}{a}$. Figure 3.7 shows the region in which $\mathbf{k}$ lies for cubic (A) and fcc lattice (B).

Then from (3.17), (3.18) and (3.19), we have for

$$C = \begin{bmatrix} K_y^2 + K_z^2 & 0 & 0 \\ 0 & K_x^2 + K_z^2 & 0 \\ 0 & 0 & K_x^2 + K_y^2 \end{bmatrix}.$$

$$\omega^2 \begin{bmatrix} A \\ B \\ C \end{bmatrix} = \underbrace{v^2 \left(2\mathbf{k}\mathbf{k}' + C\right)}_{M(\mathbf{k})} \begin{bmatrix} A \\ B \\ C \end{bmatrix}. \tag{3.23}$$

Observe $M(\mathbf{k})$ is a positive definite matrix. The three positive eigenvalues $\omega_1(\mathbf{k}), \omega_2(\mathbf{k}), \omega_3(\mathbf{k})$ give the dispersion relation for three acoustic phonons (Fig. 3.8).

3.2 Phonon Heat Capacity of Solids

Consider a cubic lattice, with phonon dispersion relation

$$\omega = v|k|. \tag{3.24}$$

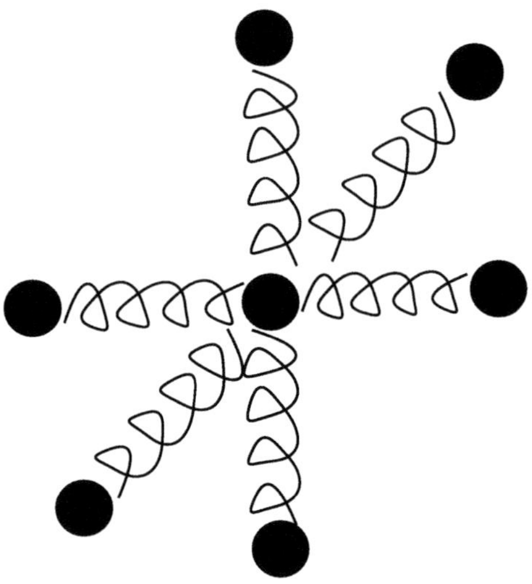

Fig. 3.8 Figure shows atomic masses and springs in three dimensions

A phonon with energy $\hbar\omega$ is active, when $\hbar\omega < k_B T$, in which case it has n quanta of energy such that $n\hbar\omega = k_B T$. To count all the phonon modes, that are active at temperature T, we have to find the volume $V = \frac{4\pi|k_0|^3}{3}$ of the sphere with radius $|k_0| = \frac{k_B T}{\hbar v}$, and since k modes are spaced $\frac{2\pi}{na}$ apart, the volume of the k cell is $V_0 = (\frac{2\pi}{na})^3$. Then the total active phonons are $\frac{V}{V_0}$, with total energy $E = k_B T \frac{V}{V_0}$. Then $\frac{dE}{dT} \propto T^3$ at low temperatures, when all phonon modes are not active. At high temperatures, when all phonon modes are active, $E \propto k_B T$ and hence heat capacity $\frac{dE}{dT}$ is independent of temperature.

3.3 Electronic Heat Capacity of Solids

To find the electron heat capacity of solids, we distinguish between two kinds of electrons, bound and mobile. The next chapter is all about this so suffice it to say that in metals and conductors, the outermost electrons of atoms are loosely bound and become mobile in a condensed state. These electrons contribute to the specific heat of a metal (Fig. 3.9).

The Fermi–Dirac distribution for occupancy of energy level with energy ϵ is

$$\phi(\epsilon) = \frac{1}{1 + \exp\left(\frac{\epsilon - \mu}{kT}\right)}. \tag{3.25}$$

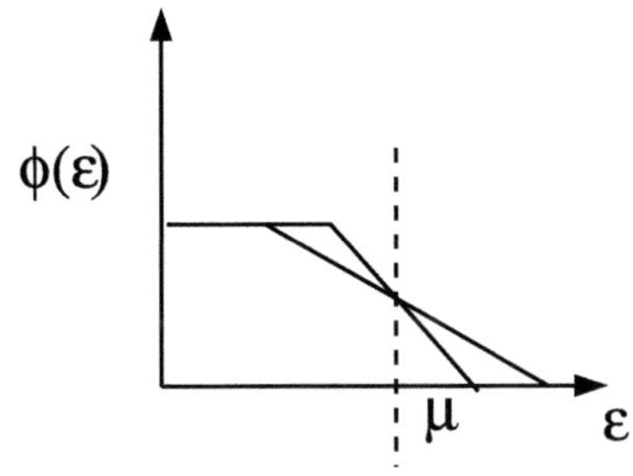

Fig. 3.9 Fermi–Dirac distribution as function of energy ϵ

For electrons in solids, the probability density of electrons near the Fermi surface μ is approximated as a linear slope. For $\Delta = \epsilon - \mu$, we can write this slope in the region $\epsilon \in [\mu - kT, \mu + kT]$

$$\phi(\epsilon) = \frac{1}{2} - \frac{\Delta}{2kT}. \tag{3.26}$$

The energy of electrons in this slope region is

$$E(T) = g(\mu) \int_{-kT}^{kT} (\mu + \Delta) \left(\frac{1}{2} - \frac{\Delta}{2kT} \right) = g(\mu) \left(\mu kT - \frac{(kT)^2}{3} \right). \tag{3.27}$$

When the temperature changes to $T + \Delta T$, this energy changes by $\Delta E = g(\mu)(\mu k - \frac{(k^2 T)}{3})\Delta T$. In addition, we lose $g(\mu)(\mu - kT)k\Delta T$. Putting the two pieces together we get

$$\Delta E = \frac{2g(\mu)}{3}(k^2 T)\Delta T.$$

Then heat capacity is $\frac{2g(\mu)kT}{3}k$. Compare with mono-atomic gas with heat capacity $\frac{3}{2}k$. We have an additional factor $\frac{2g(\mu)kT}{3}$ saying only electrons at Fermi surface contribute.

3.4 Electron–Phonon Interaction

Consider phonons in crystalline solids. To fix ideas, we start with the case of one-dimensional lattice potential. Consider a periodic potential with period a,

$$U(x) = \sum_{l=1}^{n} V(x - a_l) = \sum V(x - la),$$

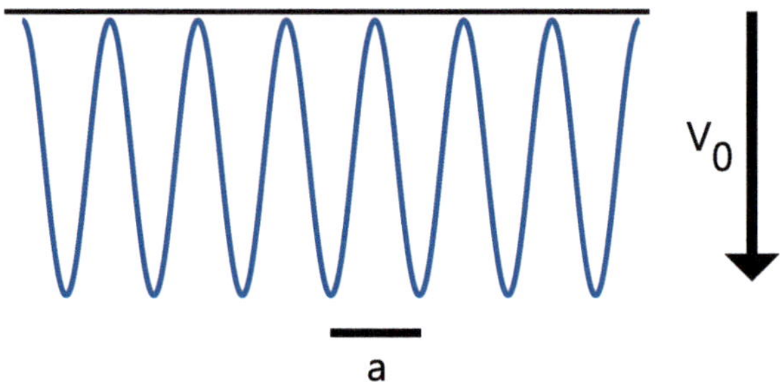

Fig. 3.10 Figure depicts the periodic potential in (3.28)

where

$$V(x) = V_0 \cos^2\left(\frac{\pi x}{a}\right), \quad \frac{-a}{2} \leq x \leq \frac{a}{2} \tag{3.28}$$

$$= 0 \quad |x| \geq \frac{a}{2}. \tag{3.29}$$

The potential is shown in Fig. 3.10.

Now consider how potential changes, when we perturb the lattice sites from their equilibrium position, due to lattice vibrations.

$$\Delta U(x) = \sum V'(x - a_l)\Delta a_l.$$

For a phonon mode with wavenumber k,

$$\Delta a_l = A_k \frac{1}{\sqrt{n}} \exp(ika_l), \tag{3.30}$$

wehave

$$\Delta U(x) = A_k \exp(ikx) \frac{1}{\sqrt{n}} \underbrace{\sum V'(x - a_l) \exp(-ik(x - a_l))}_{p(x)},$$

$$= A_k \exp(ikx) \frac{1}{\sqrt{n}} \underbrace{\sum V'(x - la) \exp(-ik(x - la))}_{p(x)},$$

where $p(x)$ is periodic with period a. Note

$$V'(x) = -V_0 \frac{2\pi}{a} \sin\left(\frac{2\pi x}{a}\right), \quad \frac{-a}{2} \leq x \leq \frac{a}{2}$$

$$= 0 \quad |x| \geq \frac{a}{2}.$$

Using Fourier series, we can write

$$p(x) = a_0 + \sum_r a_r \exp\left(i\frac{2\pi r x}{a}\right).$$

We can determine a_0 by $a_0 = \frac{1}{a}\int_{-\frac{a}{2}}^{\frac{a}{2}} p(x)dx$, giving,

$$a_0 = i\frac{V_0}{a}\int_{-\frac{a}{2}}^{\frac{a}{2}} \frac{2\pi}{a}\sin\left(\frac{2\pi x}{a}\right)\sin(kx),$$

where $k = \frac{2\pi m}{na}$. Then,

$$a_0 = i\frac{V(0)}{a}\int_{-\frac{a}{2}}^{\frac{a}{2}} \frac{2\pi}{a}\sin\left(\frac{2\pi x}{a}\right)\sin(kx) = i\frac{V(0)}{2a}\int_{-\frac{a}{2}}^{\frac{a}{2}} \frac{2\pi}{a}\left(\cos\left(\frac{2\pi q x}{a}\right) - \cos\left(\frac{2\pi q' x}{a}\right)\right),$$

where $q = 1 - \frac{m}{n}$ and $q' = 1 + \frac{m}{n}$. We then get

$$a_0 = i\frac{V(0)}{2a}\int_{-\pi}^{\pi}(\cos(qy) - \cos(q'y))dy.$$

$$a_0 = i\frac{2V(0)}{a}\frac{1}{1-\left(\frac{m}{n}\right)^2}\sin\frac{m\pi}{n}.$$

We don't worry much about a_r for $r \neq 0$, as these excite an electron to a different band and are truncated by the bandgap energy. Now note, using equipartition of energy, there is $k_B T$ energy per phonon mode, giving

$$A_k = \sqrt{\frac{k_B T}{m}\frac{1}{\omega_k}} = \sqrt{\frac{k_B T}{m}\frac{1}{\omega_d \sin\left(\frac{\pi m}{n}\right)}}, \tag{3.31}$$

where ω_d is the Debye frequency.

Then we get

$$\Delta U(x) = \frac{i}{\sqrt{n}}\underbrace{\left(\frac{2V(0)}{a}\sqrt{\frac{k_B T}{m}\frac{1}{\omega_d}}\right)}_{\bar{V}_0}\exp(ikx). \tag{3.32}$$

At temperature of $T = 300$ K and $\omega_d = 10^{13}$ rad/s, we have $\sqrt{\frac{k_B T}{m}\frac{1}{\omega_d}} \sim 0.3\,\text{A}^\circ$, with $a = 3\,\text{A}^\circ$, we have

$$\Delta U \sim \frac{i}{\sqrt{n}}0.1V_0\exp(ikx)$$

with $V_0 = 10$ V, we have

$$\Delta U \sim \frac{i}{\sqrt{n}} \exp(ikx)\ V,$$

All our analysis has been in one dimension. For higher dimensions, let's fix ideas with two dimensions, then three dimensions follow directly. Consider a two-dimensional periodic potential with period a.

$$U(x, y) = \sum_{lm} V(x - a_l, y - a_m) = \sum_{lm} V(x - la, y - ma).$$

$$V(x, y) = V_0 \cos^2\left(\frac{\pi x}{a}\right) \cos^2\left(\frac{\pi y}{a}\right), \quad \frac{-a}{2} \leq x, y \leq \frac{a}{2} \tag{3.33}$$

$$= 0 \quad |x|, |y| \geq \frac{a}{2}. \tag{3.34}$$

Now consider how potential changes when we perturb the lattice sites from their equilibrium position, due to lattice vibrations.

$$\Delta U(x, y) = \sum_{lm} V_x(x - a_l, y - a_m)\Delta a_l + V_y(x - a_l, y - a_m)\Delta a_m.$$

Let's consider phonons propagating along x direction. Then Δa_l constitutes longitudinal phonons while Δa_m constitutes transverse phonons. Transverse phonons don't contribute to deformation potential as can be seen in the following. Let's focus on the transverse phonons. Then

$$\Delta a_m = A_k \frac{1}{\sqrt{n}} \exp(ik_x a_l). \tag{3.35}$$

We have due to Δa_m

$$\Delta U(x, y) = A_k \exp(ik_x x) \frac{1}{\sqrt{n}} \underbrace{\sum V_y(x - a_l, y - a_m) \exp(-ik(x - a_l))}_{p(x,y)},$$

where $p(x, y)$ is periodic with period a.

Note

$$V_y(x, y) = -V_0 \frac{2\pi}{a} \sin\left(\frac{2\pi y}{a}\right) \cos\left(\frac{\pi x}{a}\right)^2, \quad \frac{-a}{2} \leq x, y \leq \frac{a}{2}$$

$$= 0 \quad |x|, |y| \geq \frac{a}{2}. \tag{3.36}$$

Using Fourier series, we can write

$$p(x, y) = a_0 + \sum_{r,s} a_{rs} \exp\left(i\left(\frac{2\pi r x}{a} + \frac{2\pi s y}{a}\right)\right).$$

We can determine a_0 by $a_0 = \frac{1}{a^2}\int_{-\frac{a}{2}}^{\frac{a}{2}}\int_{-\frac{a}{2}}^{\frac{a}{2}} p(x, y)dxdy$, giving $a_0 = 0$. Hence transverse phonons don't contribute. The contribution of longitudinal phonons is the same as in the 1D case.

We saw phonon $\exp(ikx)$ produces a deformation potential $\Delta U(x)$ as in (3.32). Now in three dimensions, it is

$$\Delta U(x) = \frac{i}{\sqrt{n^3}}\frac{V_0}{a}A_k(\exp(ikx)\exp(-i\omega_k t) - h.c), \tag{3.37}$$

where n^3 is number of lattice points.

Using $\frac{1}{2}M\omega_k^2 A_k^2 = n_k\hbar\omega_k$ (there are n_k quanta in the phonon), where M is the mass of ion, ω_k phonon frequency and replacing A_k we get

$$\Delta U(x) = \frac{i}{\sqrt{n^3}}\underbrace{\frac{V_0}{a}\sqrt{\frac{2\hbar}{M\omega_k}}}_{c}(\sqrt{n_k}\exp(ikx)\exp(-i\omega_k t) - h.c). \tag{3.38}$$

Thus due to phonon, electron sees a potential of form

$$\Delta U(x) = \underbrace{\frac{i\sqrt{n_k}}{\sqrt{n^3}}}_{\Omega}c(\exp(i\ (kx - \omega_k t)\ ) - \exp(-i\ (kx - \omega_k t)\ ). \tag{3.39}$$

We see this interaction potential changes the electron momentum changes by k, there must then be a corresponding change in phonon number to accommodate for the change of momentum, then writing the electron–phonon coupling Hamiltonian we get

$$H_{in} = \frac{i}{\sqrt{n^3}}c(b_k\exp(i\ (kx - \omega_k t)\ ) - b_k^\dagger\exp(-i\ (kx - \omega_k t)\ ), \tag{3.40}$$

where $b, b^\dagger$ are annihilation and creation operators for phonon.

This Hamiltonian shows coupling of electron and phonons with phonons moving, i.e., in the interaction frame of phonon Hamiltonian

$$H_p = \hbar\omega_k b_k^\dagger b_k. \tag{3.41}$$

If we include H_p, then the total Hamiltonian is

$$H_0 = H_e + H_p + H_{ep}, \tag{3.42}$$

where H_e, H_p is electron and photon Hamiltonian and H_{ep} is the coupling given by

$$H_{ep} = \frac{c}{\sqrt{n^3}} i(b \exp(ikx) - b^{\dagger} \exp(-ikx)). \tag{3.43}$$

Using a cosine potential with $V_0 \sim 10$ V, with $a \sim 3\,\text{A}^{\circ}$ and $M \sim 20$ proton masses, we have $c \sim 1$ V.

3.5 Resistance and Resonant Absorption of Phonons

Figure 3.11 shows how it is when we accelerate electrons with an electric field say in $-x$ direction. The whole Fermi sphere displaces to the right by a small amount. There is net momentum in the x direction and this constitutes the current. How does the current stop? The electrons on the right shown as black dots in Fig. 3.11 are scattered to the left as shown. This scattering is due to the absorption of phonons and annuls the forward x-momentum of electrons. How much is this scattering rate? If we absorb a phonon, the electron energy rises by $\hbar\omega_d$ and the electron scatters to states as shown in the dotted sphere as shown in Fig. 3.12.

Consider a two-level system as shown in Fig. 3.13A. Let Ω be the rate of on resonance excitation from the ground state to the excited state. The dynamics of the system with $\psi = \begin{bmatrix} \psi_1 \\ \psi_0 \end{bmatrix}$ is

$$\dot{\psi} = -\frac{i}{2} \begin{bmatrix} \omega_0 & \Omega \exp(-j\omega_0 t) \\ \Omega \exp(j\omega_0 t) & -\omega_0 \end{bmatrix} \psi. \tag{3.44}$$

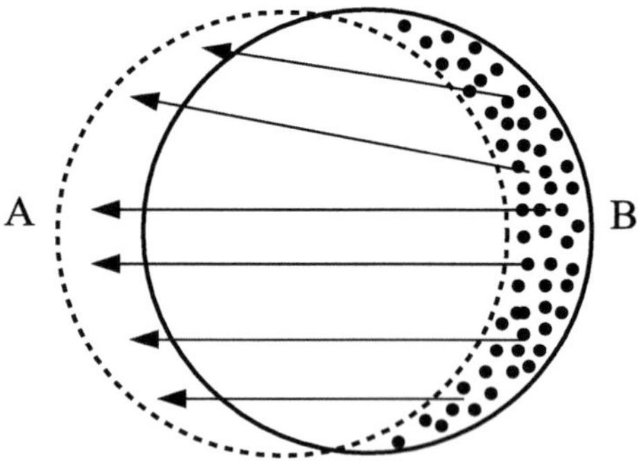

Fig. 3.11 Figure shows how it is when we accelerate electrons with an electric field say in $-x$ direction. The whole Fermi sphere displaces to the right by a small amount

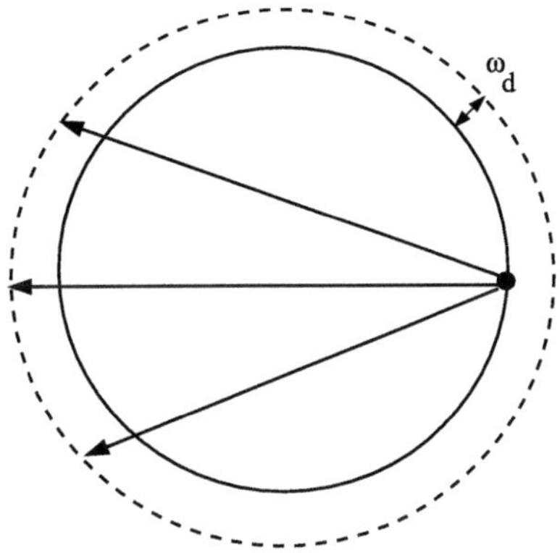

Fig. 3.12 If we absorb a phonon, the electron energy rises by $\hbar\omega_d$ and the electron scatters to states as shown in dotted sphere as shown in figure

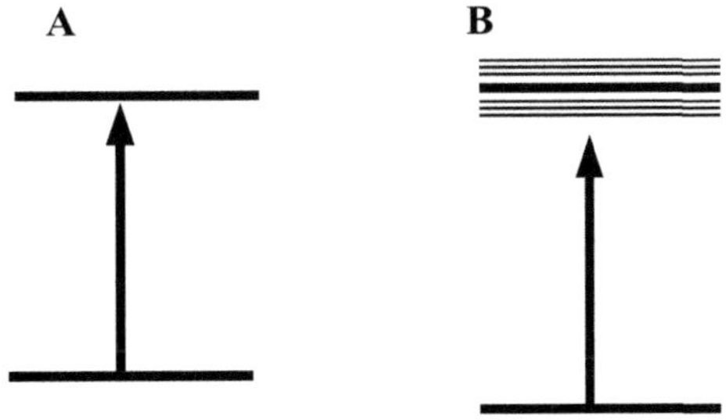

Fig. 3.13 **A** Shows transition from ground to excited state. **B** Shows excited state level broadened

In the rotating frame $\psi \rightarrow \exp\left(\frac{i}{2}\begin{bmatrix} \omega_0 & 0 \\ 0 & -\omega_0 \end{bmatrix} t\right)\psi$ we have

$$\dot{\psi} = -\frac{i}{2}\begin{bmatrix} 0 & \Omega \\ \Omega & 0 \end{bmatrix}\psi. \tag{3.45}$$

Now suppose the excited level is detuned by energy $\Delta\omega$, then

$$\dot{\psi} = -\frac{i}{2}\begin{bmatrix} 2\Delta\omega & \Omega \\ \Omega & 0 \end{bmatrix}\psi, \tag{3.46}$$

then in the rotating frame $\psi \rightarrow \exp(\frac{i}{2}\begin{bmatrix} 2\Delta\omega & 0 \\ 0 & 0 \end{bmatrix} t)\psi$, we have

$$\dot{\psi} = -\frac{i}{2}\begin{bmatrix} 0 & \Omega\exp(j\Delta\omega t) \\ \Omega\exp(-j\Delta\omega t) & 0 \end{bmatrix}\psi. \tag{3.47}$$

Then starting from $\psi_0(0) = 1$, we have $\psi_1(\Delta t) = \int_0^{\Delta t} \Omega\exp(j\Delta\omega\tau)d\tau$. Now consider many excited levels as in Fig. 3.13B, with level ψ_n detuned by $n\Delta\omega$. Then $\psi_n(\Delta t) = \int_0^{\Delta t} \Omega\exp(jn\Delta\omega\tau)d\tau$. Then

$$\psi_k^*(\Delta t)\psi_k(\Delta t) = \int_0^{\Delta t} \int_0^{\Delta t} \Omega^2 \exp(jk\Delta\omega(t-\tau))d\tau dt. \tag{3.48}$$

Then $\Phi = \sum_{k=-n}^{n} \psi_k^*(\Delta t)\psi_k(\Delta t) =$

$$\frac{\Omega^2}{\Delta\omega} \int_0^{\Delta t} \int_0^{\Delta t} \left(\int_{-B}^{B} \exp(j\omega(t-\tau))d\omega \right) d\tau dt, \tag{3.49}$$

$$= \frac{\Omega^2}{\Delta\omega} \int_0^{\Delta t} \int_0^{\Delta t} \frac{\sin(B(t-\tau)}{t-\tau} d\tau dt. \tag{3.50}$$

$$= \frac{\Omega^2}{\Delta\omega} \Delta t, \tag{3.51}$$

where we use the approximation $B \gg \Omega$. There is time Δt such that $\Omega \Delta t \ll 1$ and $B\Delta t \gg 1$. Then

$$\Gamma = \frac{\Phi}{\Delta t} = \frac{\Omega^2}{\Delta\omega}, \tag{3.52}$$

gives the transition rate out of the ground state. Let $\mathcal{N}$ be the degeneracy of every detuned state above, then

$$\Gamma = \frac{\Omega^2 \mathcal{N}}{\Delta\omega}. \tag{3.53}$$

With (3.53) also known as *Fermi Golden Rule*, we can also write it in terms of density of states $D = \frac{\mathcal{N}}{\Delta\omega}$ as

$$\Gamma = \Omega^2 D, \tag{3.54}$$

where D is density of states.

Coming to the Fig. 3.12, we can calculate Γ as follows. At say temperature $T = 100K$, we have $k_B T = 10^{-21}$ SI units. The energy of a phonon is $\hbar\omega_d \sim 10^{-21}$ SI units, so we can say that most phonon modes have one phonon.

$$\Omega = \frac{c}{\sqrt{n^3}}, \tag{3.55}$$

$$D = \frac{n^3}{\omega_F} \tag{3.56}$$

$$\Gamma = \Omega^2 D = \frac{c^2}{\omega_F}, \tag{3.57}$$

where $\hbar\omega_F$ is Fermi energy. Taking $c = 1$ eV and $\omega_F = 5$ eV, we get $\Gamma \sim 10^{14}$/s. This is in agreement with typical relaxation times of $10^{-14} - 10^{-15}$ s.

Fig. 3.14 Figure shows at low temperatures, only the phonons with small wavevectors as OA are active

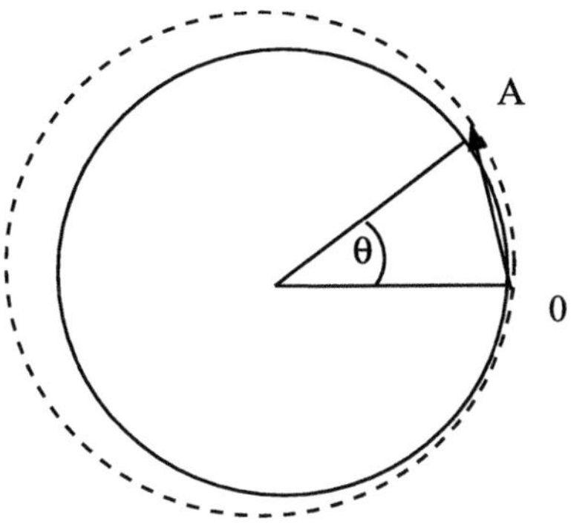

3.5.1 Temperature Dependence of Resistivity and Bloch's T^5 Law

In understanding temperature dependence of resistivity, we consider two limits. High temperature limit, when all phonon modes are occupied. Then number of phonons n_k in a mode satisfies $n_k \hbar\omega_k = k_B T$ with Ω as in (3.39), $\Omega \propto \sqrt{n_k}$. Then observe $\Omega^2 \propto k_B T$ and we have $\Gamma \propto T$. Linear variation with T.

There is another regime, the low temperature regime in which only phonons with small wavevectors which satisfy $\hbar\omega = k_B T$ are active. This is as shown by vector OA in Fig. 3.14.

The length of wavevector OA, that is active $\propto T$, and hence the surface area on the Fermi sphere, that will be active due to phonon scattering is $\propto T^2$, and hence $D \propto T^2$. Since $\Omega^2 \propto T$, we get $\Gamma = \Omega^2 D \propto T^3$. However scattering by an angle θ, as shown in Fig. 3.14, impedes the current by a factor $1 - \cos\theta$, which for small angle theta, is $\propto \theta^2$ but from Fig. 3.14, $\theta^2 \sim T^2$ or the resistivity varies as $\propto T^5$. This is called Bloch's T^5 law.

In our treatment of electron–phonon scattering, we have worked with very specific cosine lattice potential functions. However, there is genericity about our choice and the qualitative picture remains the same as working with general periodic potentials.

Problems

1. Consider a simple cubic lattice with m as the atomic mass and k as the spring constant and a as the lattice parameter. Assume interaction with the nearest and second nearest nghbs. Calculate the phonon spectrum for transverse and longitudinal phonons and sketch first Brillouin zone.
2. Consider a BCC lattice with m as atomic mass and k as spring constant and a as the lattice parameter. Assume interaction with nearest nghbs. Calculate the phonon spectrum and sketch first Brillouin zone.

3. Consider a FCC lattice with m as atomic mass and k as spring constant and a as the lattice parameter. Assume interaction with nearest nghbs. Calculate the phonon spectrum and sketch first Brillouin zone.
4. Consider a 3D periodic potential

$$V = V_0 \cos^2\left(\frac{\pi x}{a}\right)\cos^2\left(\frac{\pi y}{a}\right)\cos^2\left(\frac{\pi z}{a}\right) \tag{3.58}$$

   with $V_0 = 10$ V and lattice parameter $a = 3\,\text{A}^\circ$ and atomic mass of 20 protons. At temperature $T = 100$ K, calculate the parameter c in the book and use it to estimate the relaxation time τ assuming debye frequency of $\omega_d = 10^{13}$ Hz and Fermi energy $E_F = 10$ eV.
5. Phonon heat capacity of a solid per unit volume at low temperature goes as CT^3, find the parameter C.
6. Electron heat capacity of a solid per unit volume goes as CT, find the parameter C, assuming Fermi energy $E_F = 10$ eV.
7. Given electric field of 1 V/m and relaxation time of electrons $\tau = 10^{-14}$ s. What is the drift velocity of the electrons?

References

1. C. Kittel, *Introduction to Solid State Physics*, 8th edn. (Wiley, 2005)
2. S. Simon, *Oxford Solid State Basics* (Oxford University Press, 2013)

Chapter 4
Electrons in Periodic Potential

4.1 Free Electrons

In this chapter, we study electrons in a periodic potential, arising due to atoms in crystalline solid [1]. First, we start with free electrons.

The free electron wavefunction is $\psi = \exp(ikx)$. The momentum is $\frac{\hbar}{i}\frac{\partial}{\partial x}$. This gives the kinetic energy $\epsilon = \frac{p^2}{2m} = \frac{\hbar^2 k^2}{2m}$, which for $\epsilon = \hbar\omega$ gives

$$\omega(k) = \frac{\hbar k^2}{2m}. \tag{4.1}$$

The dispersion is a parabola, as shown in Fig. 4.1A.

Now, consider a wavepacket centered at k_0, shown the Fig. 4.1A, B. The packet takes the form

$$\phi(x) = \frac{1}{\sqrt{N}} \sum_j \exp(ik_j x), \ \ \phi(x,t) = \frac{1}{\sqrt{N}} \sum_j \exp(-i\omega(k_j)t) \exp(ik_j x), \tag{4.2}$$

where $\omega(k_j) = \omega(k_0) + \omega'(k_0)\Delta k_j$ where $\Delta k_j = k_j - k_0$. Denote $v_g = \omega'(k_0) = \frac{\hbar k_0}{m}$, as the group velocity. Then

$$\phi(x,t) = \frac{1}{\sqrt{N}} \exp\left(i(k_0 x - \omega(k_0)t)\right) \sum_j \exp(i\Delta k_j(x - v_g t)). \tag{4.3}$$

The function $f(x) = \frac{1}{\sqrt{N}} \sum_j \exp(i\Delta k_j x) = \frac{2}{\sqrt{N}} \sum_j \cos(\Delta k_j x)$, is centered at origin with width $\propto (\Delta k)^{-1}$ as shown in Fig. 4.1B. Then

$$|\phi(x,t)| = |f(x - v_g t)|, \tag{4.4}$$

N. Khaneja, *Electron Waves in Solids*, Springer Series in Solid-State Sciences 203,
https://doi.org/10.1007/978-3-031-67260-6_4

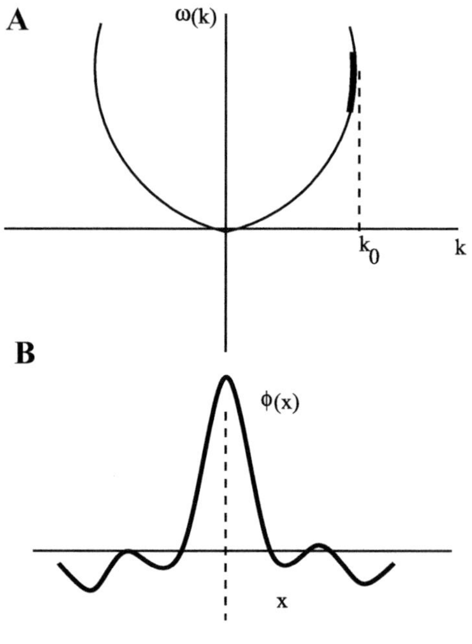

Fig. 4.1 **A** Shows the dispersion $\omega(k)$ versus k for a free electron. **B** Shows a wavepacket centered at k_0

the free electron wavepacket moves with a group velocity v_g.

Now let's apply an electrical field E in the x direction at $t = 0$. Then the Schröedinger equation is

$$i\hbar\frac{\partial\psi}{\partial t} = \frac{1}{2m}\left(- i\hbar\frac{\partial}{\partial x}\right)^2 + eEx\ \psi. \tag{4.5}$$

The equation is satisfied by time varying wavevectors $\exp(ik(t)x)$, where $k(t) = k - \frac{eEt}{\hbar}$, with energy (dispersion) $\omega(k(t)) = \frac{\hbar(k(t))^2}{2m} = \frac{\hbar(k-\frac{eEt}{\hbar})^2}{2m}$, so that the wavefunction becomes

$$\exp\left(-i\int_0^t \omega(k(\tau))\, d\tau\right)\exp(ik(t)x).$$

The initial wavepacket $\phi(x)$ evolves to $\phi(x,t)$, where

$$\phi(x) = \frac{1}{\sqrt{N}}\sum_j \exp(ik_jx),\quad \phi(x,t) = \frac{1}{\sqrt{N}}\sum_j \exp\left(-i\int_0^t \omega(k_j(\tau))d\tau\right)\exp(ik_j(t)x). \tag{4.6}$$

The group velocity

$$v_g(t) = \frac{\hbar k(t)}{m} = \frac{\hbar\left(k - \frac{eEt}{\hbar}\right)}{m};\quad \frac{dv_g(t)}{dt} = -\frac{eE}{m}. \tag{4.7}$$

The electron wavepacket simply accelerates the way we know from classical mechanics. Being more pedagogical, we have

$$\begin{aligned}\phi(x,t) &= \frac{1}{\sqrt{N}}\sum_j \exp(-i\int_0^t \omega(k_j(t))\,)\exp(ik_j(t)x) \qquad (4.8)\\ &= \frac{1}{\sqrt{N}}\exp\left(-i\int_0^t \omega(k_0(t))\right)\exp(ik_0(t)x)\sum_j \exp\left(i\Delta k_j\left(x-\int_0^t v_g(\sigma)d\sigma\right)\right).\end{aligned}$$

The wavepacket evolves with instantaneous velocity $v_g(t)$.

The above method can be generalized to arbitrary potential. Consider the Schröedinger equation

$$i\hbar\frac{\partial\psi}{\partial t} = \left(-\frac{\hbar^2}{2m}\frac{\partial}{\partial x^2} - eV(x)\right)\psi. \qquad (4.9)$$

We approximate the potential V by piecewise linear potential such that $V(x) = V(x_i) + V'(x_i)\delta x$, where $\delta x = x - x_i$, as shown in Fig. 4.2. We call these regions of linearized potential, cells. We can rewrite the potential in a cell as $V(x) = U(x_i) + V'(x_i)x$.

We assume that the wavepacket has large k_0 such that $\Delta k \sim \sqrt{k_0}$ is large and therefore for the wavepacket, $\Delta x \sim (\Delta k)^{-1}$ is small so that it fits well within one cell. Then in this cell, the Schröedinger equation takes the form

$$i\hbar\frac{\partial\psi}{\partial t} = \left(-\frac{\hbar^2}{2m}\frac{\partial}{\partial x^2} - eV'(x_i)x - eU(x_i)\right)\psi. \qquad (4.10)$$

Since the wavepacket is confined to a cell, its evolution would be the same if the potential we have was not only true in the cell but globally true. This is because the wavepacket doesn't know what the potential is outside the cell, it's confined to the cell. Then let's solve the Schröedinger equation with this potential assumed globally true and see how wavepacket evolves.

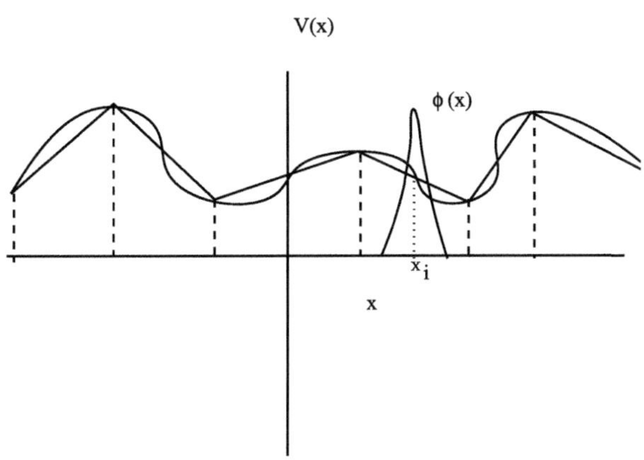

Fig. 4.2 Figure shows linear approximation of potential $V(x)$. The wavepacket $\phi(x)$ is confined to a cell

Then as before for the Schröedinger equation is solved by wavevector $\psi = \exp(ik(t)x)$. Let $x(t)$ denote coordinates of the center of wavepacket, then

$$k(t) = k + \frac{e\int_0^t V'(x(\tau))d\tau}{\hbar}, \quad \omega(k(t)) = \frac{\hbar\left(k + \frac{e\int_0^t V'(x(\tau))d\tau}{\hbar}\right)^2}{2m} - \frac{e\int_0^t U(x(\tau))d\tau}{\hbar}. \tag{4.11}$$

The group velocity

$$v_g(t) = \frac{\hbar\left(k + \frac{e\int_0^t V'(x(\tau))d\tau}{\hbar}\right)}{m}; \quad \frac{dv_g(t)}{dt} = \frac{eV'(x(t))}{m}. \tag{4.12}$$

This is classical mechanics. Therefore at high energies, where k_0 is large, and wavepacket is well confined, i.e., over the packet width, the second order change of potential is small, $V''(x)\Delta x \ll V'(x)$. A linearized potential is a good approximation and evolution in quantum mechanics mimics classical mechanics.

4.2 Electrons in Periodic Potential

We now turn to electrons in a periodic potential, created by the periodic arrangement of ions in crystal/solid. Once again to ease exposition, we first analyze everything in one dimension.

Consider a periodic potential with period a.

$$U(x) = \sum_{l=1}^{n} V(x - a_l) = \sum V(x - la),$$

where we choose a model potential $V(x)$, which captures the essence of the problem.

$$V(x) = 2V_0\cos^2\left(\frac{\pi x}{a}\right), \quad \frac{-a}{2} \leq x \leq \frac{a}{2} \tag{4.13}$$

$$= 0 \quad |x| \geq \frac{a}{2}. \tag{4.14}$$

The potential is shown in Fig. 4.3.

How does the wavepacket evolve in this potential? This potential is rapidly varying on length scale of a (few A°). To treat the particle classically, the packet width should be confined to $\Delta x \sim a$. This gives $\Delta k = a^{-1}$. This means that packet should have $k_0 \gg a^{-1}$. However at room temperature, using $\frac{mv^2}{2} = \frac{3kT}{2}$, we get de Broglie wavelength $\lambda_0 = 2\pi k_0^{-1} = \frac{h}{p}$ of 100's of A°. Hence electron is much delocalized compared to the scale at which potential varies and we have to treat the problem of the evolution of electron in the periodic potential quantum mechanically. Let's see how we do this.

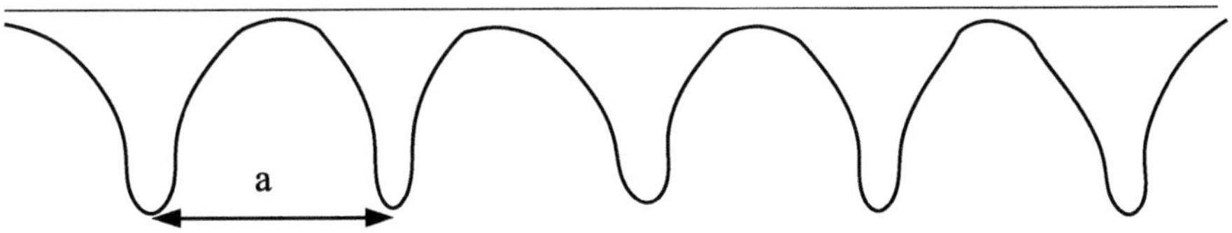

Fig. 4.3 Figure depicts the periodic potential in (4.13)

Let a be the lattice constant such that periodic potential $V(x) = V(x+a)$. We can Fourier decompose

$$V(x) = \sum_k V_k \exp(ikx). \quad V_{-k} = V_k^*.$$

For example when $V(x)$ is as in (4.13), we have

$$V(x) = V_0 \left(1 + \frac{\exp(i\frac{2\pi}{a})}{2} + \frac{\exp(-i\frac{2\pi}{a})}{2}\right).$$

What does this potential do? It couples wavevectors $k \pm \frac{2\pi m}{a}$. Then suppose we have a wavefunction of the form

$$\phi(x) = \sum_m b_m \exp\left(i\left(k + \frac{2\pi m}{a}\right)x\right)$$

then we can write the Schröedinger equation as

$$\frac{d}{dt}\begin{bmatrix} \vdots \\ b_{-1} \\ b_0 \\ b_1 \\ \vdots \end{bmatrix} = -iH \begin{bmatrix} \vdots \\ b_{-1} \\ b_0 \\ b_1 \\ \vdots \end{bmatrix},$$

where

$$H = \begin{bmatrix} \ddots & \cdots & \cdots & \cdots & \cdots \\ 0 & \frac{\hbar^2(k+\frac{2\pi}{a})^2}{2m} + V_0 & \frac{V_0}{2} & \cdots & 0 \\ 0 & \frac{V_0}{2} & \frac{\hbar^2 k^2}{2m} + V_0 & \frac{V_0}{2} & 0 \\ 0 & 0 & \frac{V_0}{2} & \frac{\hbar^2(k-\frac{2\pi}{a})^2}{2m} + V_0 & \frac{V_0}{2} \\ 0 & \cdots & \cdots & \cdots & \ddots \end{bmatrix}. \tag{4.15}$$

The V_0 on the diagonal is just a phase factor, we can remove it to get

$$H = \begin{bmatrix} \ddots & \cdots & \cdots & \cdots & \cdots \\ 0 & \frac{\hbar^2(k+\frac{2\pi}{a})^2}{2m} & \frac{V_0}{2} & \cdots & 0 \\ 0 & \frac{V_0}{2} & \frac{\hbar^2 k^2}{2m} & \frac{V_0}{2} & 0 \\ 0 & 0 & \frac{V_0}{2} & \frac{\hbar^2(k-\frac{2\pi}{a})^2}{2m} & \frac{V_0}{2} \\ 0 & \cdots & \cdots & \cdots & \ddots \end{bmatrix}. \tag{4.16}$$

The eigenfunctions of the Bloch electrons take the general form

$$\psi_k(x) = \exp(ikx)u_{k,n}(x),$$

where k is the crystal momentum lying in the range $\frac{-\pi}{a} \leq k \leq \frac{\pi}{a}$, $u_{k,n}(x+a) = u_{k,n}(x)$ is the periodic part of the wavefunction and n is the band index. We have

$$u_{k,n}(x) = \sum_k b_k \exp\left(-i\frac{2\pi kx}{a}\right).$$

The corresponding energies $\omega_{k,n}$ are eigenvalues of the system Hamiltonian

$$H = \begin{bmatrix} \ddots & \cdots & \cdots & \cdots & \cdots \\ 0 & \frac{\hbar^2(k+\frac{2\pi}{a})^2}{2m} + V_0 & \frac{V_0}{2} & \cdots & 0 \\ 0 & \frac{V_0}{2} & \frac{\hbar^2 k^2}{2m} + V_0 & \frac{V_0}{2} & 0 \\ 0 & 0 & \frac{V_0}{2} & \frac{\hbar^2(k-\frac{2\pi}{a})^2}{2m} + V_0 & \frac{V_0}{2} \\ 0 & \cdots & \cdots & \cdots & \ddots \end{bmatrix}, \tag{4.17}$$

which for different l, couples free electron states $\exp(i(k + \frac{2\pi l}{a}))$, with the periodic potential.

If $(\ldots, b_1, b_0, b_{-1}, \ldots)'$ is the eigenvector of H in (4.17) corresponding to energy $\epsilon_{k,n} = \hbar\omega_{k,n}$, then

$$u_{k,n}(x) = \sum_l b_l \exp\left(i\frac{2\pi l}{a}x\right); \quad \psi_{k,n}(x) = \sum_l b_l \exp\left(i\left(k + \frac{2\pi l}{a}\right)x\right).$$

$$H = \begin{bmatrix} \ddots & \cdots & \cdots & \cdots & \cdots \\ 0 & \frac{\hbar^2(k+\frac{2\pi}{a})^2}{2m} & \frac{V_0}{2} & \cdots & 0 \\ 0 & \frac{V_0}{2} & \frac{\hbar^2 k^2}{2m} & \frac{V_0}{2} & 0 \\ 0 & 0 & \frac{V_0}{2} & \frac{\hbar^2(k-\frac{2\pi}{a})^2}{2m} & \frac{V_0}{2} \\ 0 & \cdots & \cdots & \cdots & \ddots \end{bmatrix}. \tag{4.18}$$

Let's us focus on a two by two block as shown below. For $k \in [0, \frac{\pi}{a}]$, the remaining matrix in (4.17) can be taken to be diagonal as off-diagonal entries are truncated.

$$H = \begin{bmatrix} \frac{\hbar^2 k^2}{2m} & \frac{V_0}{2} \\ \frac{V_0}{2} & \frac{\hbar^2 (k - \frac{2\pi}{a})^2}{2m} \end{bmatrix}. \tag{4.19}$$

In Fig. 4.4, we sketch $\omega(k)$ as function of k (right half).

The eigenvalues of (4.19) can be written in closed form. For

$$H = \begin{bmatrix} \mu_1 & V_0 \\ V_0 & \mu_2 \end{bmatrix}, \tag{4.20}$$

we have the eigenvalues

$$\frac{\mu_1 + \mu_2}{2} \pm \sqrt{\left(\frac{\mu_1 - \mu_2}{2}\right)^2 + V_0^2}.$$

At $k = \frac{\pi}{a}$, we have $\mu_1 = \mu_2$ and the difference of eigenvalues is called the bandgap.

For $k \in [-\frac{\pi}{a}, 0]$, we focus on the following sub-block of (4.17).

$$H = \begin{bmatrix} \frac{\hbar^2 k^2}{2m} & V_0 \\ V_0 & \frac{\hbar^2 (k + \frac{2\pi}{a})^2}{2m} \end{bmatrix}. \tag{4.21}$$

This helps us sketch $\omega(k)$ as function of k (left half).

We now consider when only $n = 0$ is populated, we suppress band index from now with $n = 0$ implied. Now consider a wavepacket around $k = k_0$ written as

$$\phi(x) = \frac{1}{\sqrt{N}} \sum_j \psi_{k_j}(x) = \frac{1}{\sqrt{N}} u(x) \sum_j \exp(i k_j x),$$

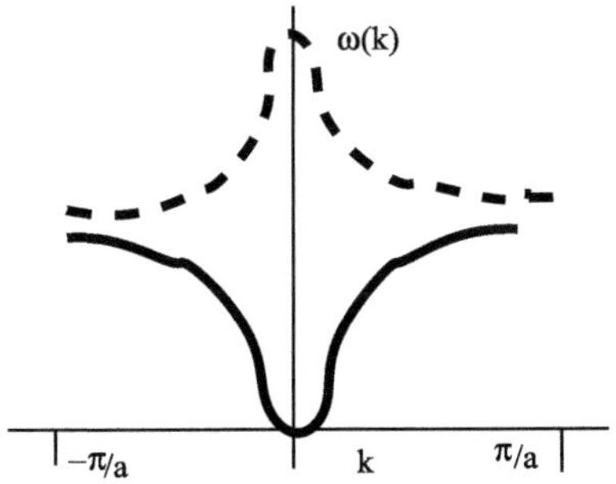

Fig. 4.4 Figure depicts the energy ω_k as function of k for periodic potential in (4.19)

where $u_k(x)$ for k around k_0 are approximated to be same $u(x)$. Then this wavepacket evolves as

$$\begin{aligned}\phi(x,t) &= \frac{u(x)}{\sqrt{N}}\sum_j \exp(-i\omega(k_j)t)\exp(ik_jx) \\ &= \frac{u(x)}{\sqrt{N}}\exp(-i\omega(k_0)t)\exp(ik_0x)\sum_j \exp(i\Delta k_j(x-v_gt)),\end{aligned}$$

where $v_g = \frac{d\omega}{dk}|_{k_0}$. The wavepacket evolves with velocity v_g.

Now let's analyze the evolution of the wavepacket in the presence of an electric field E, say in x direction. We again introduce time varying $k(t)$ with

$$k(t) = k + \frac{eA(t)}{\hbar}, \quad A(t) = -Et,$$

then

$$H(t) = \begin{bmatrix} \ddots & \cdots & \cdots & \cdots & \cdots \\ 0 & \frac{\hbar^2(k+\frac{2\pi}{a}+\frac{eA(t)}{\hbar})^2}{2m}+V_0 & \frac{V_0}{2} & \cdots & 0 \\ 0 & \frac{V_0}{2} & \frac{\hbar^2(k+\frac{eA(t)}{\hbar})^2}{2m}+V_0 & \frac{V_0}{2} & 0 \\ 0 & 0 & \frac{V_0}{2} & \frac{\hbar^2(k-\frac{2\pi}{a}+\frac{eA(t)}{\hbar})^2}{2m}+V_0 & \frac{V_0}{2} \\ 0 & \cdots & \cdots & \cdots & \ddots \end{bmatrix}. \tag{4.22}$$

Then observe

$$\omega_{k,n}(t) = \omega_{k+\frac{eA(t)}{\hbar},n} = \omega_{k-\frac{eEt}{\hbar},n}. \tag{4.23}$$

This is how energies of $H(t)$ change when we apply electric field. To solve for time varying Schröedinger equation we have to realize that for moderate E, the change of $H(t)$ is adiabatic and hence we really just follow the eigenvectors of $H(t)$. To see how this adiabatic following works, say at $t = 0$, we are in the ground state say $X(0)$ (note $X(0) = (\ldots, a_1, a_0, a_{-1}, \ldots)'$, where $\psi_k(x) = \sum_l a_l \exp(i(k+\frac{2\pi l}{a})x)$.) then $X(t)$ satisfies the Schröedinger equation

$$\dot{X} = \frac{-i}{\hbar}H(t)X.$$

Let's diagonalize $H(t)$ as

$$H(t) = \Theta(t)\Lambda(t)\Theta'(t),$$

where $\Theta(t)$ is matrix of eigenvectors and $\Lambda(t)$ eigenvalues. $H(t)$ is symmetric and $\Theta(t)$ a real rotation matrix. As we will see, this will ensure we don't have any geometric phases in our adiabatic evolution. Then we have $\dot{\Theta}(t) = \Omega(t)\Theta$, where $\Omega(t)$ is a skew symmetric matrix and we get for $Y(t) = \Theta'(t)X(t)$,

$$\dot{Y} = \left(\frac{-i}{\hbar} \begin{bmatrix} \lambda_1 & 0 & 0 & 0 \\ 0 & \lambda_2 & 0 & 0 \\ \vdots & \dots & \ddots & \vdots \\ 0 & 0 & 0 & \lambda_n \end{bmatrix} + \underbrace{\Theta'\Omega(t)\Theta}_{\bar{\Omega}} \right) Y, \tag{4.24}$$

where λ_1 is the smallest eigenvalue and so on. Note $Y(0) = (1, 0, \dots, 0)'$.

Now there is a gap between the lowest eigenvalue and higher eigenvalues, which is at least the bandgap in the range of eV. $\bar{\Omega}_{1j}(t)$ are comparatively small as we will show in following, and therefore they average out and $Y(t)$ simply evolves at $Y(t) = \exp(-\frac{i}{\hbar}\int \lambda_1(\sigma))Y(0) = \exp(-i\int \omega_k(\sigma))Y(0)$ and we get that we just adiabatically follow the eigenvector of H to get

$$\psi(t) = \exp(-i\int \omega_k(\sigma)d\sigma)\psi_{k+\frac{eA(t)}{\hbar}} = \exp\left(-i\int \omega_{k-\frac{eE\sigma}{\hbar}} d\sigma\right) u_{k-\frac{eEt}{\hbar}}(x)\exp(ik(t)x).$$

To estimate how big are elements $\bar{\Omega}_{1j}(t)$, observe

$$\begin{aligned} \dot{H} &= \Theta(t)\dot{\Lambda}(t)\Theta'(t) + [\Omega, H] \\ \Theta'(t)\dot{H}\Theta(t) &= \dot{\Lambda}(t) + [\bar{\Omega}, \Lambda] \\ \|[\bar{\Omega}, \Lambda]\| &\leq \|\Theta'(t)\dot{H}\Theta(t)\| = \|\dot{H}\|. \end{aligned}$$

Let $\bar{\Omega}_*(t) = \max\{\bar{\Omega}_{1j}(t)\}$, then we have

$$(\lambda_1 - \lambda_2)\bar{\Omega}_*(t) \leq \|[\bar{\Omega}, \Lambda]\| \leq \|\dot{H}\|. \tag{4.25}$$

Note we truncate H in (4.17) to finite size, because when l on diagonal terms of H become large the corresponding off-diagonals are rapidly truncated and in computing the ground state the sub-matrix with few small l suffices. Usually $n = 10$ is appropriate when V_l are of order eV.

When we compute $\dot{H}$ we only have terms on diagonal and they are of the form

$$\frac{eE\hbar}{m}\left(k + \frac{eA(t)}{\hbar}\right),$$

then

$$\|\dot{H}\| = \frac{eE}{\sqrt{m}}\sqrt{\sum_l \frac{\hbar^2(k + \frac{2\pi l}{a} + \frac{eA(t)}{\hbar})^2}{m}} = \frac{eE}{\sqrt{m}}\sqrt{2tr(\bar{H})}, \tag{4.26}$$

where $\bar{H}$ is just H without V_0 on the diagonal. Then $\bar{\Omega}_*(t)$ from (4.25) satisfies

$$\hbar\bar{\Omega}_*(t) \leq \frac{\hbar e E}{\sqrt{m}(\lambda_1 - \lambda_2)}\sqrt{2tr(\bar{H})}. \tag{4.27}$$

Observe $(\lambda_1 - \lambda_2)$ is few eV and $tr(\bar{H})$ is also tens of eV, but for field say $E =$ 10 V/m, the term

$$\frac{\hbar e E}{\sqrt{m}} \sim \frac{10^{-33} \times 10^{-18}}{10^{-15}} = 10^{-36}.$$

$$\frac{\hbar e E}{\sqrt{m}(\lambda_1 - \lambda_2)}\sqrt{tr(\bar{H})} = 10^{-36}/\sqrt{10^{-18}} = 10^{-27}\,\text{J} = 10^{-8}\,\text{eV}.$$

Now we observe that $(\lambda_1 - \lambda_2)$ is of order eV while $\hbar\bar{\Omega}_*(t)$ is of order 10^{-8} eV, then in (4.24), we have elements $\bar{\Omega}_{1l}$ average out, giving us a pure adiabatic evolution.

It should be noted that in the equation

$$\omega_k(t) = \omega_{k-\frac{eEt}{\hbar}},$$

we treat $k - \frac{eEt}{\hbar}$ in reduced Brillouin zone, i.e, for $b > 0, k - \frac{eEt}{\hbar} = \frac{\pi}{a} + b$ is reduced to $-\frac{\pi}{a} + b \in [-\frac{\pi}{a}, \frac{\pi}{a}]$.

Now let's envisage a situation where we turn on an electric field in a conductor for some time τ and switch it off. Then $A(t) = -\int_0^t E(\tau)d\tau$ rises from 0 to a steady value A. This can be achieved as shown in Fig. 4.5, where a conducting loop is pierced by a solenoid of area a_r. When current in the solenoid is turned on from 0 to a steady state value, that creates a magnetic field $B(t)$ which goes from 0 to a steady state

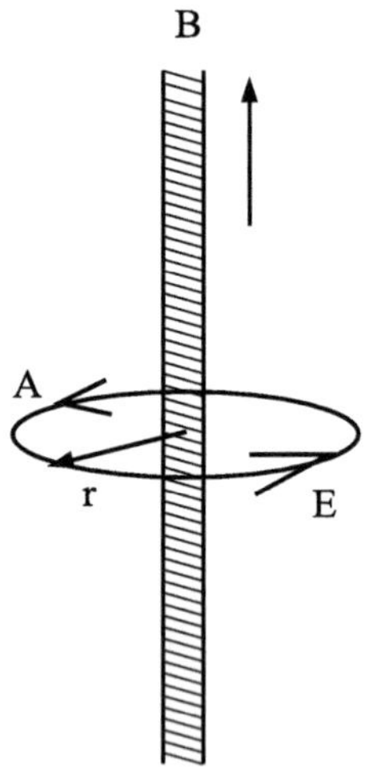

Fig. 4.5 Figure shows a solenoid inside a loop conductor. As current in the solenoid is turned on a magnetic field B is established inside the solenoid, which creates a E in the loop that rises and decays and A that goes from 0 to a steady state value

value B and hence establishes a transient electric field in the conducting loop and finally results in a steady A in the loop such that

$$A = \frac{Ba_r}{2\pi r}.$$

Then the electron wavefunction $\psi_k = \exp(ikx)u_k(x)$ is adiabatically transformed to $\exp(i(k + \frac{eA}{\hbar})x)u_{k+\frac{eA}{\hbar}}(x)$. The initial wavepacket $\phi(x) = \frac{1}{\sqrt{N}}u_{k_0}(x)\sum_j \exp(k_j x)$ is adiabatically transformed to

$$\phi(x) = \frac{1}{\sqrt{N}}u_{k_0+\frac{eA}{\hbar}}(x)\sum_j \exp\left(\left(k_j + \frac{eA}{\hbar}\right)x\right).$$

If initially the wavepacket was moving with a group velocity $v_g = \frac{d\omega(k)}{dk}|_{k_0}$, now it moves with a group velocity $v_g = \frac{d\omega(k)}{dk}|_{k_0+\frac{eA}{\hbar}}$. This is shown in Fig. 4.6, where $E(t)$ is transiently turned on and off and $A(t)$ reaches a steady state value A. The figure shows how A in the conducting loop shifts the energy of a wavepacket and thereby changes the group velocity and hence accelerates the wavepacket. Let's estimate the shift $\frac{eA}{\hbar}$ for the example we have. Suppose solenoid has $N = 10$ turns per c.m. and carries a final current of $I = 1$ milli Ampere. Then it establishes a B field of $B = \mu_0 NI$. Let r be 0.1 m and $a_r = 1\,\text{cm}^2$. Then we have $\frac{eA}{\hbar} = 10^4\,\text{m}^{-1}$. At half filling $k_0 = \frac{\pi}{2a}$, with $a = 1\text{A}^\circ$, $k_0 \sim 10^{10}\,\text{m}^{-1}$ and is displaced 0.0001% by electric field.

We can now understand conduction in solids. Consider a half filled band in a conductor (say the conducting loop in the above example) as shown in Fig. 4.7a. There is no net velocity of electrons. There are as many wavepackets moving in the positive direction as negative direction. While the wavepacket centered at k_0 has group velocity $v_g = \frac{d\omega}{dk}|_{k_0}$, there is a wavepacket at $-k_0$ with group velocity $-v_g$. There is no net current, though electrons are itinerant. Now say we apply a transient electric field as in solenoid example above which shifts the energy $\omega_k \to \omega_{k+\frac{eA}{\hbar}}$. Then we reach a configuration as shown in Fig. 4.7b. Then we see we have an excess of wavepackets with positive v_g. In the solenoid example we calculated the excess to be around 0.0001%. This gives net conduction and current. Therefore half filled bands conduct. Now imagine an insulator where the band is completely filled, then all $k \in [-\frac{\pi}{a}, \frac{\pi}{a}]$ are occupied. After application of E we have $\omega_k \to \omega_{k+\frac{eA}{\hbar}}$, but because we are in the reduced Brillouin zone, nothing happens, as all energies are simply shifted, the energy of one wavepacket takes the value of another and so on. Although individual wavepackets are accelerated and deaccelerated the sum total of velocities is still zero. It is a common understanding that in an insulator all $k's$ are filled so the electric field cannot accelerate a wavepacket and everything is stuck. In our picture, wavepackets are accelerated/deaccelerated but the sum total of v_g of all the wavepackets remains zero.

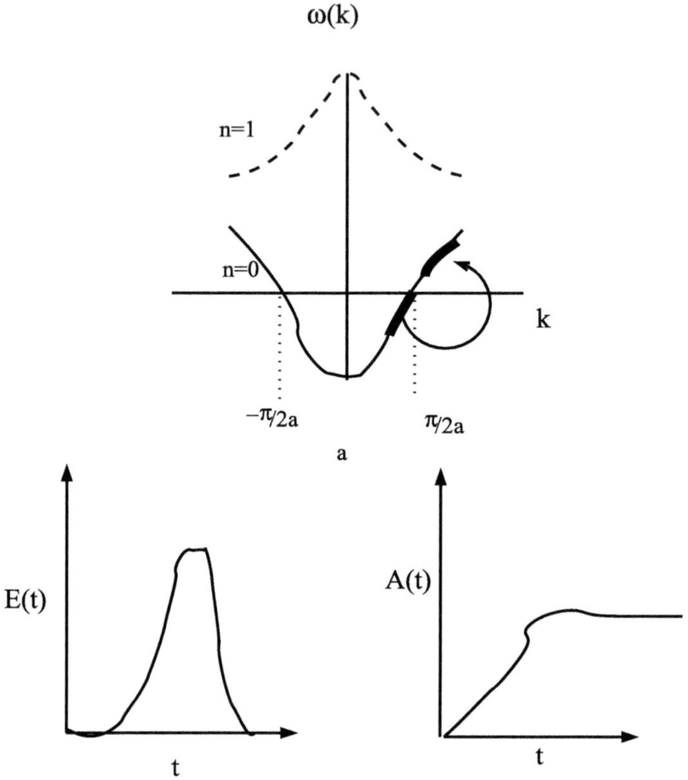

Fig. 4.6 Bottom figure shows how application of current in solenoid in Fig. 4.5 establishes a transient E and A in the conducting loop which shifts the energy of a wavepacket and thereby changes the group velocity and hence accelerates the wavepacket as in top figure

In an insulator as the electron wavefunction $\psi_k = \exp(ikx)u_k(x)$ is adiabatically transformed to $\exp(i(k + \frac{eA}{\hbar})x)u_{k+\frac{eA}{\hbar}}(x)$, when $k + \frac{eA}{\hbar} > \frac{\pi}{a}$, then putting it back in reduced Brillouin zone let $k + \frac{eA}{\hbar} = \frac{2\pi}{a} - k'$, with $k' \in [\frac{-\pi}{a}, \frac{\pi}{a}]$, then observe $u_{k+\frac{eA}{\hbar}}(x)$ really means $\exp(-i\frac{2\pi}{a}x)u_{-k'}(x)$.

Now in the presence of an electric field, we have

$$\omega_k(t) = \omega_{k-\frac{eEt}{\hbar}}. \tag{4.28}$$

Then the group velocity satisfies

$$v_g(t) = \frac{d\omega_{k-\frac{eEt}{\hbar}}}{dk}, \tag{4.29}$$

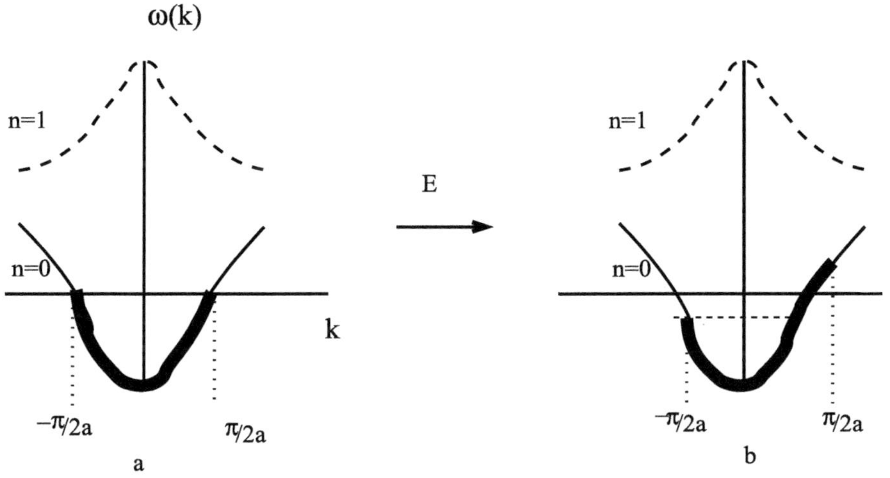

Fig. 4.7 **a** Shows a half filled band. **b** Shows how by application of electric field, the energies are shifted

$$\frac{dv_g(t)}{dt} = \frac{d}{dk}\frac{d\omega_{k-\frac{eEt}{\hbar}}}{dt} = \frac{d^2\omega_{k-\frac{eEt}{\hbar}}}{dk^2}\frac{eE}{\hbar} = \hbar^{-2}\frac{d^2\epsilon_{k-\frac{eEt}{\hbar}}}{dk^2}eE$$
$$= -\frac{1}{m^*}eE, \tag{4.30}$$

where $m^* = \hbar^2\left(\frac{d^2\epsilon_{k+\frac{eEt}{\hbar}}}{dk^2}\right)^{-1}$ is the effective mass.

For pedagogical reasons we have restricted to 1D, we can easily generalize the above to 3D. Let ∇' be gradient written as a column vector and ∇ be gradient written as a row vector. Writing v_g as column 3 vector,

$$v_g(t) = \nabla'_k\omega_{k+\frac{eEt}{\hbar}}, \tag{4.31}$$

$$\frac{dv_g(t)}{dt} = \nabla'_k\frac{d\omega_{k-\frac{eEt}{\hbar}}}{dt} = \nabla'_k\nabla_k\omega_{k-\frac{eEt}{\hbar}}\frac{eE}{\hbar} = \hbar^{-2}\nabla'_k\nabla_k\epsilon_{k-\frac{eEt}{\hbar}}eE$$
$$= -\frac{1}{m^*}eE, \tag{4.32}$$

where $m^* = \hbar^2(\nabla'_k\nabla_k\epsilon_{k+\frac{eEt}{\hbar}})^{-1}$ is the effective mass matrix.

We derived wavepacket dynamics of an electron in periodic potential. Of course, the whole treatment in this paper is in the absence of a resistance. The wavepacket in reality scatters of phonons and impurities. The present treatment only details the dynamics of electron wavepacket between collisions with lattice (Fig. 4.8).

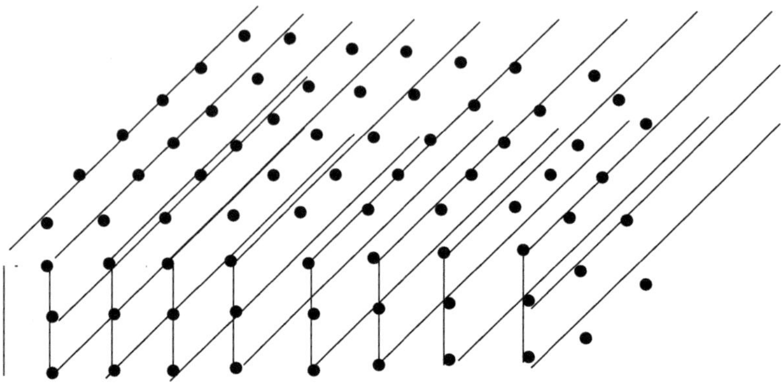

Fig. 4.8 Figure shows a 3D lattice made from periodic arrangement of positively charged ions

4.3 Periodic Potential in Three Dimensions: Real Solids

In last section, we studied one-dimensional periodic lattices and showed how electron energies get organized as bands, where each band is made up of block vectors with their wavevector taking on values between $[\frac{-\pi}{a}, \frac{-\pi}{a}]$. All this can be generalized to three dimensions. We can talk about plain wave states $\exp(-i(k_x x + k_y y + k_z z))$, which we abbreviate (k_x, k_y, k_z), for the electrons and how periodic potential couples them. For example, periodic potential can be simplified to

$$V(x, y, z) = 8V_0 \cos^2\left(\frac{\pi x}{a}\right) \cos^2\left(\frac{\pi y}{a}\right) \cos^2\left(\frac{\pi z}{a}\right), \tag{4.33}$$

which is a periodic potential in three dimensions, with period a. The potential will couple plain wave states $(k_x \pm \frac{2\pi l}{a}, k_y \pm \frac{2\pi m}{a}, k_z \pm \frac{2\pi n}{a})$. Let

$$E(k_x, k_y, k_z) = \frac{\hbar^2(k_x^2 + k_y^2 + k_z^2)}{2m}.$$

Then for each (k_x, k_y, k_z), we can form the matrix

$$H = \begin{bmatrix} \ddots & \cdots & \cdots & \cdots & \cdots \\ 0 & E(k_x + \frac{2\pi}{a}, k_y, k_z) & \frac{V_0}{2} & \cdots & 0 \\ 0 & \frac{V_0}{2} & E(k_x, k_y, k_z) & \frac{V_0}{2} & 0 \\ 0 & 0 & \frac{V_0}{2} & E(k_x - \frac{2\pi}{a}, k_y, k_z) & \frac{V_0}{2} \\ 0 & \cdots & \cdots & \cdots & \ddots \end{bmatrix}, \tag{4.34}$$

when we diagonalize this matrix, we get the energies $\epsilon_n(k_x, k_y, k_z)$, where n indices the various bands. The corresponding eigenvectors are

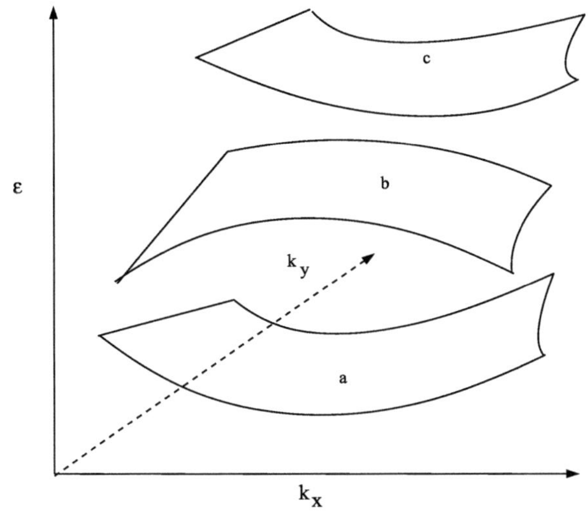

Fig. 4.9 Figure shows energy bands for periodic two-dimensional lattice

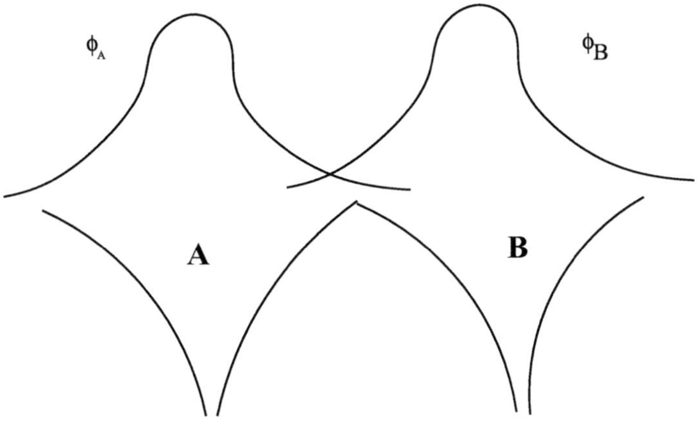

Fig. 4.10 Figure shows two potential wells **A** and **B** and their corresponding orbitals with overlap between them

$$u_n(k_x, k_y, k_z) = \sum_{lmn} b_{lmn} \left(k_x \pm \frac{2\pi l}{a}, k_y \pm \frac{2\pi m}{a}, k_z \pm \frac{2\pi n}{a} \right),$$

we do it for all $-\frac{\pi}{a} \leq k_x, k_y, k_z \leq \frac{\pi}{a}$ and obtain energy bands. Figure 4.10 shows energy bands for periodic lattice in two dimensions (Fig. 4.9).

The approach we have described in this section is so-called the nearly free electron approximation. We start with free electrons and couple them with lattice potential. There is another method for solving for electron energies and wavefunctions in a periodic potential. It is called the tight-binding approximation. Here our starting eigenvectors are atomic orbitals localized on atom sites. As we bring atoms together in the lattice, these orbitals interact with the potential of neighboring ions and get delocalized. We can solve for these delocalized waves and their energies as shown in the following section.

4.4 Tight Binding Approximation

The basic idea is as follows. Consider a periodic array of say hydrogen atoms. If we take an isolated hydrogen atom then the electron around its nucleus organizes as orbitals with energies ϵ_k. Now suppose we bring hydrogen atoms together, as long as they are far nothing changes, as each hydrogen atom has its own electron and they are isolated because the electron of one hydrogen atom sees no potential due to another because the electron in the other atom screens its charge. Then if we have an array of isolated hydrogen atoms, well separated then the energies and eigenvectors are as for individual ones. But now suppose we bring the atoms further closer so that the electron of atom A enters the electron cloud of electron B. Then the electron of atom A will begin to also see some potential due to ion B, as its charge is not fully screened and this will modify the wavefunction and energies of electron A.

Let us call the potentials of ions V_A and V_B and the wavefunctions ϕ_A and ϕ_B each with energy ϵ_0. There is a transition from ϕ_A to ϕ_B given by

$$\langle \phi_B V_B \phi_A \rangle = -t$$

and similarly by symmetry there is transition from ϕ_B to ϕ_A given by

$$\langle \phi_A V_A \phi_B \rangle = -t.$$

Then in the basis ϕ_A and ϕ_B, we can write the Hamiltonian as

$$H = \begin{bmatrix} \epsilon_0 & -t \\ -t & \epsilon_0 \end{bmatrix}. \tag{4.35}$$

The eigenvalues now are $\epsilon_0 \mp t$ and eigenvectors $\frac{1}{2}(\phi_A \pm \phi_B)$. The two electrons will now occupy the lower energy orbital $\frac{1}{2}(\phi_A + \phi_B)$. The new orbitals are a linear combination of atomic orbital (LCAO) (Fig. 4.11).

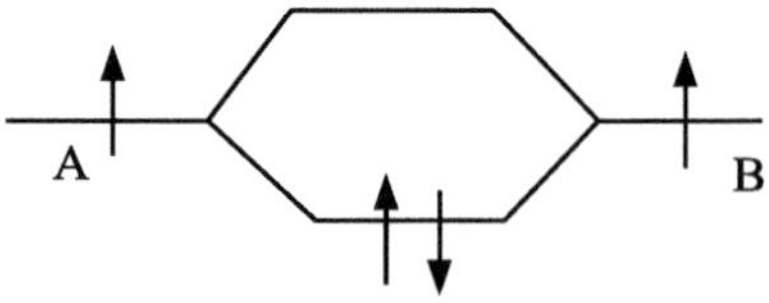

Fig. 4.11 Figure shows two atomic orbitals ϕ_A and ϕ_B and their linear combinations with energy $\epsilon_0 - t$ and $\epsilon_0 + t$

Thus we have seen how atomic orbitals are modified when we bring atoms closer. We form molecular orbitals. Now we extend this to a periodic array say first in one dimension, also called a 1 d chain.

$$H = \begin{bmatrix} \epsilon_0 & -t & 0 & \dots\dots & & -t \\ -t & \epsilon_0 & -t & 0 & \ddots & \vdots \\ 0 & -t & \epsilon_0 & -t & 0 & \vdots \\ \vdots & \ddots & \ddots & \ddots & \ddots & \vdots \\ 0 & \dots\dots & & -t & \epsilon_0 & -t \\ -t & 0 & \dots\dots & & -t & \epsilon_0 \end{bmatrix}, \tag{4.36}$$

where we have introduced a transition between 1 and n to close the chain. H above is a circulant matrix of the form

$$\begin{bmatrix} r_0 & r_1 & r_2 & \dots\dots & & r_n \\ r_n & r_0 & r_1 & r_2 & \ddots & \vdots \\ r_{n-1} & r_n & r_0 & r_1 & r_2 & \vdots \\ \vdots & \ddots & \ddots & \ddots & \ddots & \vdots \\ r_2 & \dots\dots & & r_n & r_0 & r_1 \\ r_1 & r_2 & \dots\dots & & r_n & r_0 \end{bmatrix}. \tag{4.37}$$

A circulant matrix is always diaganolized by a DFT matrix whose jth column is $\frac{1}{\sqrt{n}}\begin{pmatrix} 1 \\ \omega^{j-1} \\ \omega^{2(j-1)} \\ \dots \\ \omega^{(n-1)(j-1)} \end{pmatrix}$ with eigenvalue $\sum_k r_k(\omega^{j-1})^{k-1}$ where ω in nth root of unity $\omega = \exp(i\frac{2\pi}{n})$. Then our eigenvalues are

$$\epsilon_0 - 2t\cos\frac{2\pi j}{n}, \quad 0 < j < n-1,$$

which can be written as

$$\epsilon(k) = \epsilon_0 - 2t\cos ka, \quad \frac{\pi}{a} < k < \frac{\pi}{a}. \tag{4.38}$$

This is called a dispersion relation. The eigen vectors are of the form

$$\Psi(x) = \sum_l \exp(ikla)\phi(x - la). \tag{4.39}$$

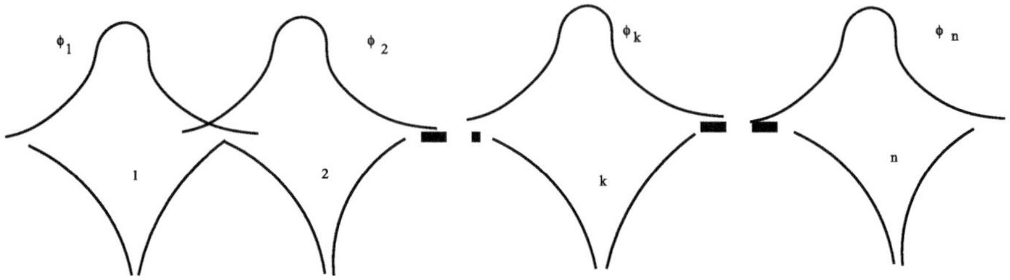

Fig. 4.12 Figure shows atomic orbitals ϕ_1 to ϕ_n and their potential wells

Recall the eigenfunction for the Block electrons in a periodic potential. It has the form $\exp(ikx)p(x)$ where $p(x)$ is periodic with period a. The orbital in (4.39) is similar, except its exponential part is sampled at lattice points (Fig. 4.12).

Thus starting from isolated atomic orbitals, with one energy ϵ_0, we branch into many energies as we decrease the spacing between atoms. The energy gets broadened into a band. This is shown in Fig. 4.13.

Now coming back to 1 d chain of hydrogen-like atoms, we found that given N atoms, we get N wavevectors with their k ranging from $\frac{-\pi}{a}$ to $\frac{\pi}{a}$, we have N orbitals with wavevector separation of $\frac{2\pi}{N}$ between them. These are shown in Fig. 4.14. By Pauli exclusion principle, each orbital can carry only 2 electrons with opposite spins. Since we have N electrons, if we fill the energy levels starting from minimum energy, we get only $\frac{N}{2}$ filled orbitals as each can hold two electrons. The remaining $\frac{N}{2}$ orbitals are empty. When we discussed chain of periodic potentials, we showed how half filled bands conduct. Same is true now. After we do tight-binding approximation, we still have the same principle, half filled bands conduct. We can form electron wavepackets

$$\frac{1}{\sqrt{n}} \sum_j \exp(ik_j x)p(x) \tag{4.40}$$

Fig. 4.13 Figure shows how atomic orbital energy ϵ_0 gets branched into many energies as we decrease the separation a between orbitals

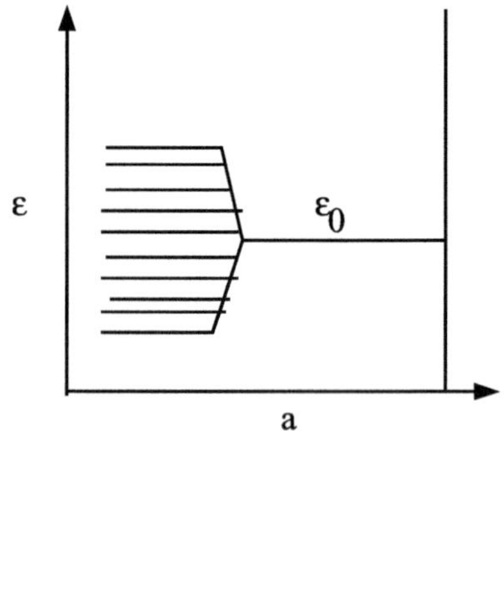

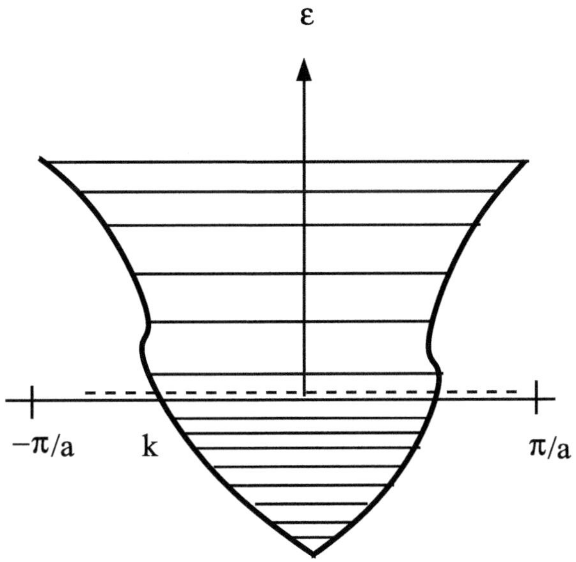

Fig. 4.14 Figure shows a plot of dispersion relation in (4.38)

centered at wavevector k_0 with $p(x)$ the atomic or periodic part. Then this will evolve with group velocity $\hbar^{-1}\frac{d\epsilon(k)}{dk}|_{k_0}$. In the presence of electric field, it evolves as

$$\frac{1}{\sqrt{n}}\sum_j \exp(ik_j(t)x)p(x), \tag{4.41}$$

where $k_j(t) = k_j - \frac{eEt}{\hbar}$, with center $k_0(t) = k_0 - \frac{eEt}{\hbar}$. Then this will evolve with group velocity

$$v_g(t) = \hbar^{-1}\frac{d\epsilon(k)}{dk}|_{k_0(t)}. \tag{4.42}$$

Observe then

$$\frac{dv_g(t)}{dt} = \hbar^{-1}\frac{d^2\epsilon(k)}{dk^2}|_{k_0(t)}(\dot{k}_0(t)) = -\underbrace{\hbar^{-2}\frac{d^2\epsilon(k)}{dk^2}|_{k_0(t)}}_{1/m^*}eE. \tag{4.43}$$

The wavepacket moves like a classical particle in an electric field E with mass m^*.

Now imagine an atom with many orbitals with energies $\epsilon_1, \ldots, \epsilon_N$. Then as we bring these atoms together, these energies will get broadened as shown in Fig. 4.15.

The tight-binding approximation has been described using a 1D chain. In real solids, we have a periodic arrangement of ions in three dimensions. Then we have to solve tight-binding approximation for three dimensions. To do this let us recall

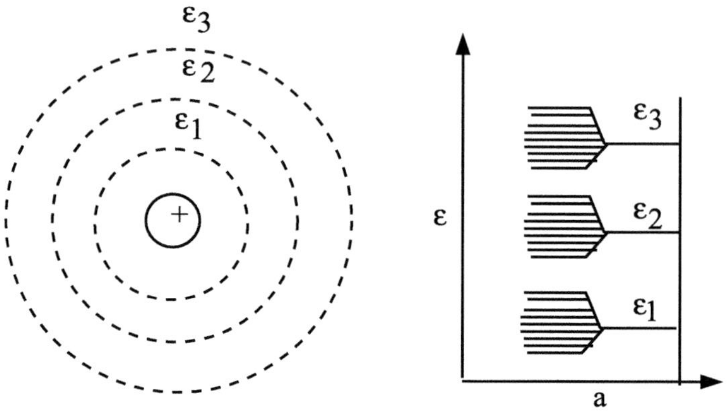

Fig. 4.15 Figure shows how many orbitals with energies $\epsilon_1, \ldots, \epsilon_N$ get broadened as we decrease the separation a between orbitals

1D in a different format. There we solve a eigenvalue problem of the form (x_l are eigenvector entries)

$$-t(x_{l-1} + x_{l+1}) = \epsilon x_l. \tag{4.44}$$

This is solved by using a wave-ansatz $x_l = \exp(ikla) = \omega^l$ with $\epsilon = -2t\cos(ka)$, with $\frac{-\pi}{a} \leq k \leq \frac{\pi}{a}$.

How does this problem look in two dimensions with a periodic arrangement of ions with spacing a. We find we have equation,

$$-t(x_{l-1,m} + x_{l+1,m} + x_{l,m-1} + x_{l,m+1}) = \epsilon x_{lm}. \tag{4.45}$$

This is solved by using a wave-ansatz $x_{lm} = \exp(ik_x la)\exp(ik_y ma)$ with $\epsilon = -2t(\cos(k_x a) + \cos(k_y b))$ with $\frac{-\pi}{a} \leq k_x, k_y \leq \frac{\pi}{a}$. We plot below the function $\epsilon(k_x, k_y)$ (Fig. 4.16).

How does this problem look in three dimensions with a periodic arrangement of ions with spacing a. We find we have an equation,

$$-t(x_{l-1,m,n} + x_{l+1,m,n} + x_{l,m-1,n} + x_{l,m+1,n} + x_{l,m,n-1} + x_{l,m,n+1}) = \epsilon x_{l,m,n}. \tag{4.46}$$

This is solved by using a wave-ansatz $x_{lmn} = \exp(ik_x la)\exp(ik_y ma)\exp(ik_z na)$ with $\epsilon = -2t(\cos(k_x a) + \cos(k_y a) + \cos(k_z a))$ with $\frac{-\pi}{a} \leq k_x, k_y, k_z \leq \frac{\pi}{a}$.

In the following Fig. 4.17, we show how does energies of atomic orbitals in a 3D solid say (sodium or calcium) get broadened as we bring the atoms closer. For sodium (atomic number 11) we have electronic state $1s^2 2s^2 2p^6 3s^1$ with only 1 electrons in $3s$. As this energy level gets broadened we have two electrons in each k-state giving that the band is half filled. As a result Sodium is a conductor. How about calcium (atomic number 12), it has electronic state $1s^2$, $2s^2$, $2p^6$, $3s^2$ and has

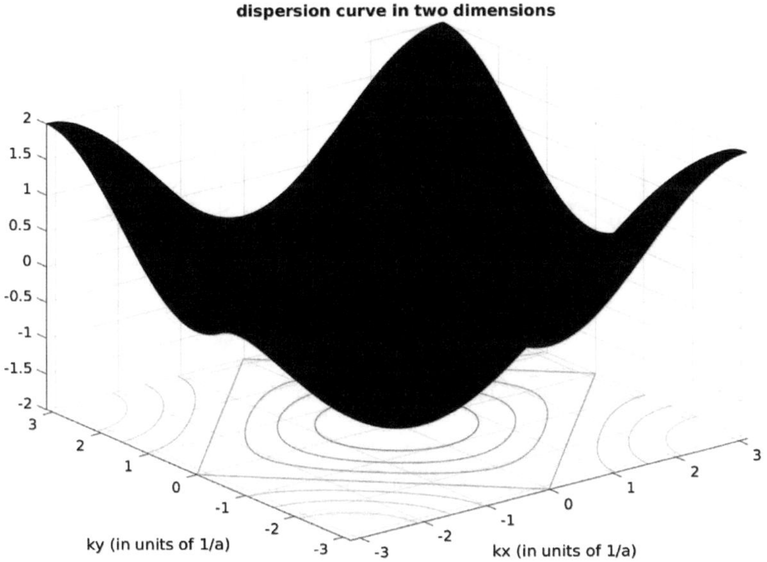

Fig. 4.16 Figure shows a plot of $\epsilon = -2t(\cos(k_x a) + \cos(k_y b))$ as function of (k_x, k_y)

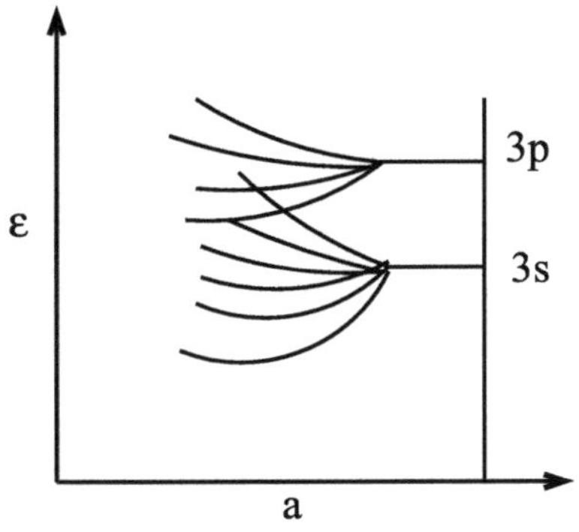

Fig. 4.17 Figure shows how 3s and 3p orbitals get broadened for an atom like sodium or calcium

2 electrons in $3s$. As energy level gets broadened, we have two electrons in each k-state, giving that the s band is fully filled hence we should have calcium as an insulator. However s band overlaps with p band, and hence both of then get partially filled and we get a conductor. For Aluminum (atomic number 13) we have electronic state $1s^2 2s^2 2p^6 3s^2 3p^1$ and we have three electrons. Again we fill both s and p and the s band may get fully filled but p band is partially filled and we have a conductor. Hence all these are metals, they are good conductors.

Now let's take the next element silicon (atomic number 14) we have electronic state $1s^2 2s^2 2p^6 3s^2 3p^2$. Since silicon will occupy most of the remaining chapter, we spend some time studying its band structure. Its band structure is very interesting.

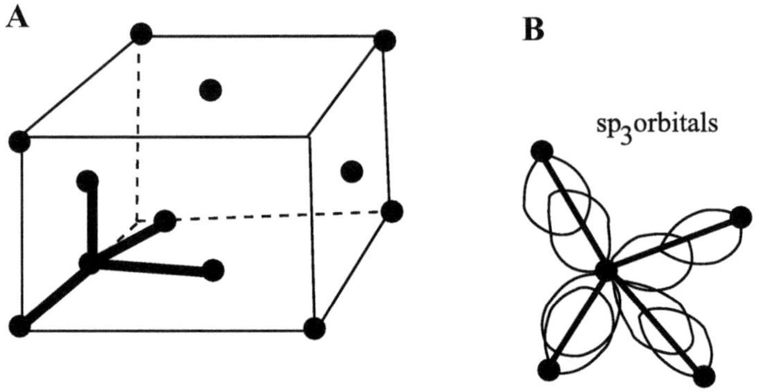

Fig. 4.18 **A** Shows two inter-spaced fcc lattices that make the silicon lattice

There are atoms at the corner of the cube and at the center of all faces of a cube. Such a structure is called the fcc structure. Silicon lattice is made of such a fcc lattice and its displaced version by an amount $\frac{a}{4}(1, 1, 1)$ as shown in Fig. 4.18B. There is 1 3s orbital, and three 3p orbitals, which combine to form four sp^3 orbitals. Each silicon atom say at lattice position $\frac{a}{4}(1, 1, 1)$ will form bonds with its four immediate nghs $(0, 0, 0)$, $a(\frac{1}{2}, \frac{1}{2}, 0)$, $a(0, \frac{1}{2}, \frac{1}{2})$ and $a(\frac{1}{2}, 0, \frac{1}{2})$ with one sp^3 orbital each, as shown in Fig. 4.18B. The two sp^3 orbitals of the neighboring silicon atoms strongly overlap as shown in Fig. 4.18B. Then as described in Fig. 4.12, the two sp^3 atomic orbitals combine to form two molecular orbitals at energies $\epsilon + t$ and $\epsilon - t$. Now these molecular orbitals also overlap, and hence we have a picture like following Fig. 4.19, where molecular orbitals are formed from atomic orbitals as in Fig. 4.19a, which due to overlap between them get further broadened as in Fig. 4.19b with Δ as the gap between the lower and higher manifold as shown in Fig. 4.19b.

We develop a simple model to illustrate this process of first forming molecular orbitals from atomic orbitals and then broadening them. The sp^3 orbital has two lobes a big positive lobe and a small negative lobe as shown in Fig. 4.22A. Consider a linear chain of atoms with two kinds of orbitals at each site ϕ_i and $\bar{\phi}_i$. With transfer integral $-t_1$ between ϕ_i and $\bar{\phi}_{i+1}$ and transfer integral t_2 between ϕ_i and ϕ_{i+1} (positive lobe of ϕ_i talks to negative lobe of ϕ_{i+1}). This is shown in Fig. 4.20. Then we have molecular orbitals $\Phi_i = \frac{\phi_i + \bar{\phi}_{i+1}}{\sqrt{2}}$ with energy $\epsilon - t_1$ and $\bar{\Phi}_i = \frac{\phi_i - \bar{\phi}_{i+1}}{\sqrt{2}}$ with energy $\epsilon + t_1$. But now Φ_i talk to each other with transfer integral t_2 between them and Φ_i' talk to each other with transfer integral t_2 between them. Then Φ_i gets broadened into a band and Φ_i' get broadened into another band as shown in Fig. 4.21.

Observe the maximum of the lower energy manifold (called valence band) and the minimum of the higher energy manifold (called conduction band) happens at different k. Hence this is an indirect bandgap. This is basically what happens in a silicon crystal. If we have two electrons per atomic orbital (left and right as in Fig. 4.20) then all the lower energy manifold made out of Φ_i will be filled with two electrons per orbital and we will end up in an insulator. However, if the bandgap is

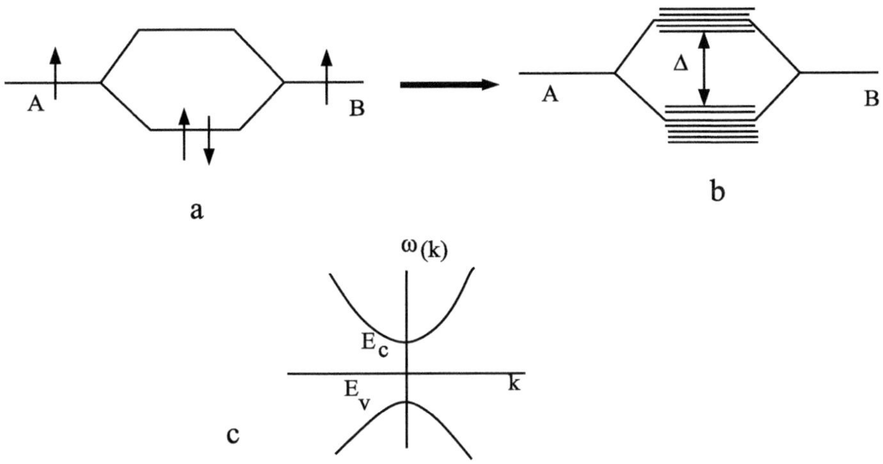

Fig. 4.19 Figure shows molecular orbitals are formed from atomic orbitals as in (**a**) which due to overlap between them get further broadened as in (**b**). **c** Shows conduction and valence band. The energy at bottom of conduction band is E_c and at top of valence band is E_V

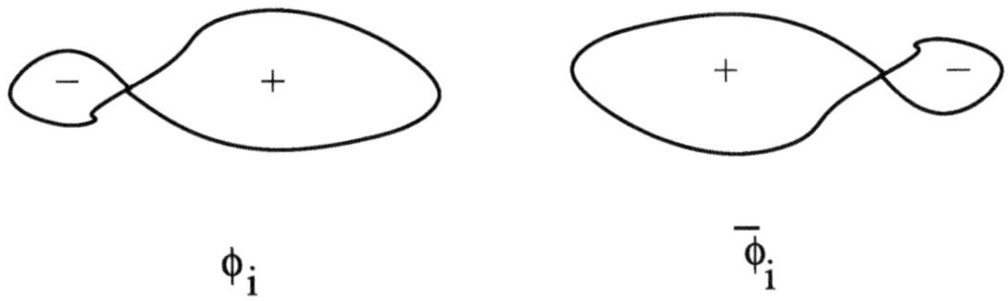

Fig. 4.20 Figure shows two kind of atomic orbitals ϕ facing right and $\bar{\phi}$ facing left

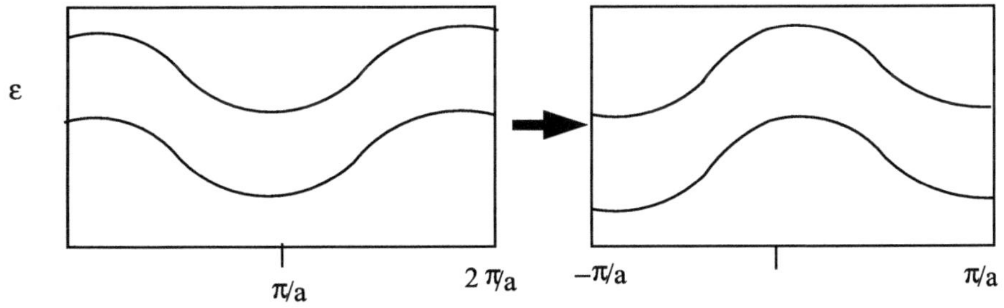

Fig. 4.21 The figure shows indirect bandgap for sp hybridized orbitals in a model for silicon

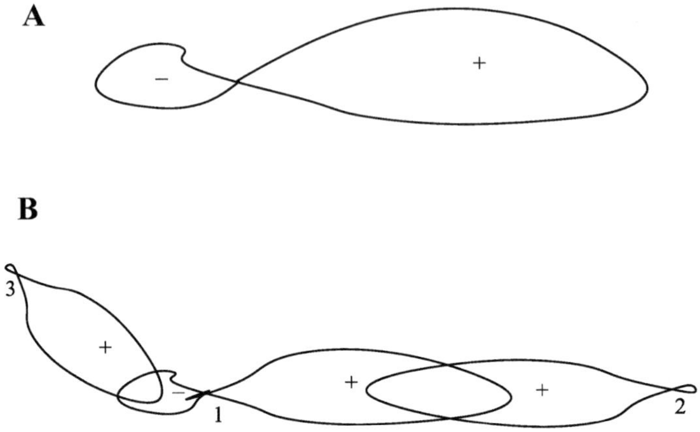

Fig. 4.22 **A** Shows two lobes of a sp^3 orbital. **B** Shows how these sp^3 orbitals overlap in a crystal

not too high (bandgap in silicon is around 1.1 eV) some of the electrons from the valence band will be excited to the conduction band at high temperatures. This will make the material a conductor. Such a material which is insulator at low temperatures and conductor at high is called a semiconductor. Silicon is a prime example.

We now look at another semiconductor, Gallium Arsenide (GaAs). Its structure is same as silicon as shown in Fig. 4.18A, except now we have a sites of Ga and b sites of As. Outer electronic configuration of Ga is $4s^2 4p^1$ and As is $4s^2 4p^3$. Like silicon they also form sp^3 orbitals except now sp^3 orbitals of As are more electronegative and have lower energy compared to Ga. Then a sites have energy $\epsilon_1 = \epsilon_0 + \Delta\epsilon$ and b sites have energy $\epsilon_2 = \epsilon_0 - \Delta\epsilon$. Although we have transfer integral between orbitals on a and b sites, due to large energy difference between the two sites, it gets truncated, instead between consecutive a sites we have a second order coupling between ϕ_i and $\bar{\phi}_{i+2}$ of $-t_1'$ and between ϕ_i and ϕ_{i+2} of t_2'. Similarly between consecutive b sites we have a second order coupling between ϕ_i and $\bar{\phi}_{i+2}$ of t_1' and between ϕ_i and ϕ_{i+2} of $-t_2'$. Then b sites give two bands as in Fig. 4.21 at lower energies (of b sites) and also two bands at higher energies (of a sites). The lower two bands are the valence band that holds $2n$ electrons (Fig. 4.23).

To understand what this second order coupling is let $H = \begin{bmatrix} \epsilon_1 & t & 0 \\ t & \epsilon_2 & t \\ 0 & t & \epsilon_1 \end{bmatrix}$ the coupling Hamiltonian of three site system $a - b - a$, with ϵ_1 energy of site a and ϵ_2 energy of site b. t is the coupling between the sites. The 13 entry in this matrix is zero, but if we evolve H, we get an effective 13 coupling as a second order term of magnitude $\frac{t^2}{\epsilon_1 - \epsilon_2}$ and furthermore due to large energy difference $\epsilon_1 - \epsilon_2$ the 12 and 23 entries are suppressed in effective evolution.

As remarked before in a semiconductor at low temperatures valence band is full and the conduction band is empty, as we raise the temperature electrons from the top

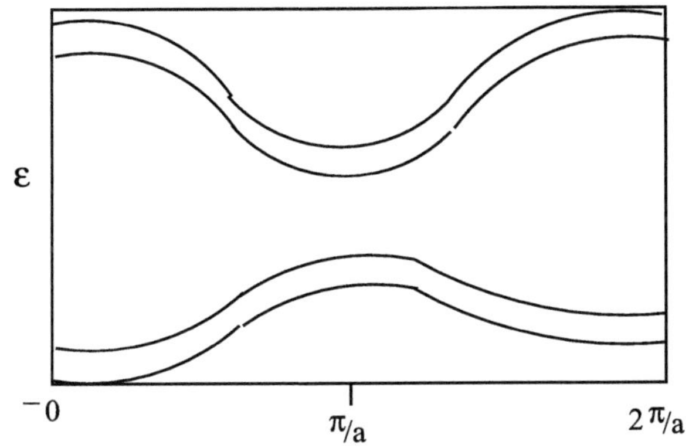

Fig. 4.23 The figure shows direct bandgap for material like GaAs

of the valence band are excited to the conduction band. These excited electrons may fall back and the difference of energy of the bands may be emitted as light, given be relation $\hbar\omega = \Delta E$. The wavenumber of this light $k = \frac{\omega}{c}$. For ΔE in couple of eV, we have ω as 10^{14} and corresponds to $k = 10^6$. Compared to wavenumber $\frac{\pi}{a}$, this is negligible. Hence conservation of momentum says that the wavenumber of electron in conduction and valence band should be the same, which is true in direct bandgap materials. Hence direct bandgap materials can emit light. For instance, GaAs emit red light. If we take GaP (Gallium Phosphide) then the bandgap is even bigger, as phosphide is more electronegative than As and hence ΔE is bigger. Giving light of higher frequency, i.e., green light. It is even more in GaN (Gallium Nitride), giving blue light. With these three colors, we can make white light which is how our LED bulb works. We will say more about this when we discuss Light emitting diodes (LED).

Problems

1. Consider a simple cubic lattice with a and t as the lattice parameter and the hopping parameter and onsite energy ϵ_0. Find dispersion relation $\epsilon(k)$, where assume interaction with nearest nghbs. Assuming one electron per site, how is the band filled?
2. Consider a BCC lattice with a and t as the lattice parameter and the hopping parameter and onsite energy ϵ_0. Find dispersion relation $\epsilon(k)$. where assume interaction with nearest nghbs. Assuming one electron per site how is the band filled?
3. Consider an FCC lattice with a and t as the lattice parameter and the hopping parameter and onsite energy ϵ_0. Find dispersion relation $\epsilon(k)$, where assume interaction with nearest nghbs. Assuming one electron per site how is the band filled?

4. Consider a Silicon lattice with a and t as the lattice parameter and the hopping parameter and onsite energy ϵ_0. Assuming four sp^3 hybridized electron per site, find dispersion relation $\epsilon(k)$ for valence and conduction band.
5. Consider a Gallium Arsenide lattice with a and t as the lattice parameter and the hopping parameter and onsite energy ϵ_1 on Gallium and ϵ_2 at Arsenic. Assuming four sp^3 hybridized electron per site find dispersion relation $\epsilon(k)$ for valence and conduction band.
6. Consider a 1D periodic potential with lattice parameter $a = 3\,\text{A}^\circ$ and t parameter at 5 eV. Calculate the fermi velocity of a half filled band.
7. Now consider a 3D periodic potential with simple cubic lattice and lattice parameter $a = 3\,\text{A}^\circ$ and t parameter at 5 eV. Calculate the fermi velocity for an electron in half filled band on Fermi sphere in direction $(1, 0, 0)$.
8. Consider a 1D periodic potential with electrons treated in free electron approximation. With potential

$$V = V_0 \cos^2\left(\frac{\pi x}{a}\right) \tag{4.47}$$

with $V_0 = 100\,\text{V}$ and lattice parameter $a = 3\,\text{A}^\circ$. Sketch first two energy bands and find the bandgap.
9. In the above find the bandgap between 2nd and 3rd band.
10. Consider a 3D periodic potential with electrons treated in free electron approximation. With potential

$$V = V_0 \cos^2\left(\frac{\pi x}{a}\right)\cos^2\left(\frac{\pi y}{a}\right)\cos^2\left(\frac{\pi z}{a}\right), \tag{4.48}$$

with $V_0 = 100\,\text{V}$ and lattice parameter $a = 3\,\text{A}^\circ$. find the bandgap between first two energy levels.

Reference

1. J. Singleton, *Band Theory and Electronic Properties of Solids* (Oxford University Press, 2001)

Chapter 5
Electronic Devices

5.1 Doping

Let's consider silicon again, and now suppose, we replace some of them with phosphorus atoms. Silicon outer shell is $3s^2 3p^2$, while in phosphorus we have $3s^2 3p^3$. Thus phosphorus will give one more electron. We may just say that in silicon valence band is all filled, so the extra electrons given by the phosphorus atom will just go into the conduction band, making silicon a conductor. We say silicon is n-doped. We talked about n doped silicon. We can replace silicon with Aluminum Al, whose outer shell is $3s^2 3p^1$. Not all valence bands will be filled now, there are vacancies, we call *holes*. Since we have a partially filled band, we will again get a conductor. This is called a p doped silicon.

Most of modern electronics [1, 2] are playing with n and p doped silicon, and making useful devices out of them. The next few sections are devoted to the description of these devices. We start with the most basic of these called a p-n junction.

5.2 p-n Junction

Figure 5.1a shows a pn junction with p-doped silicon on the right and n doped silicon on the left. When we apply a positive voltage to p junction compared to n junction (forward bias), the following happens. The valence electron in the p region moves toward positive terminal, creating a positive p region, which pulls the conduction electrons of the n region into the p region, and current flows, see Fig. 5.1b. Some of these electrons arriving from the n-region will recombine with the holes in the valence band. Nonetheless, there is a current flow in the forward bias.

Now apply a positive voltage to the n junction, compared to p junction (reverse bias), the following happens. We cannot pull the valence of p, they are filled. The valence electrons in p region cannot move in n junction, as all valence states in n

N. Khaneja, *Electron Waves in Solids*, Springer Series in Solid-State Sciences 203,
https://doi.org/10.1007/978-3-031-67260-6_5

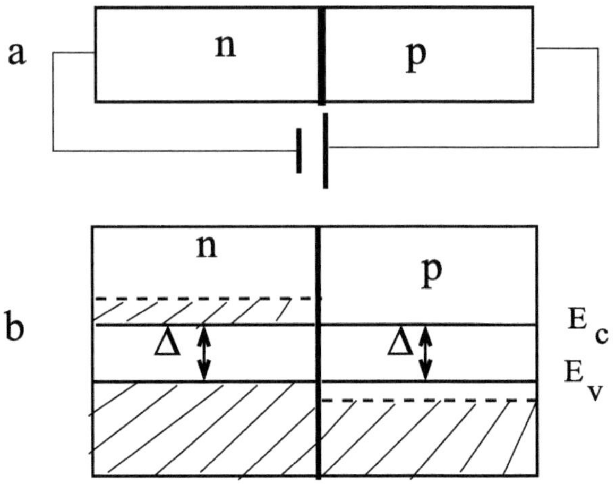

Fig. 5.1 **a** Shows a pn junction. **b** Shows conduction and valence band separated by energy gap Δ. Dotted lines show the fermi level, the energy up to which the band is full (at 0 temperature). E_c and E_v are bottom and top of conduction and valence band energies respectively

region are filled and they do not move under an electric field, see Fig. 5.1b. Thus we do not have current flow in reverse bias.

In practice, when we bring a p and n junction in contact there is a transfer of charge between them. The conduction electrons in n region are at higher energy, compared to electrons in the p region, which are in the valence band. Hence there is the flow of electrons from n region to p region. This leaves n region positively charged, and p region negatively charged, and an electric field develops at the interface as shown in Fig. 5.2A, which prevents further flow of charges. Under this condition, if we make p terminal slightly positive, the current won't flow. The n region conduction electrons are attracted to p region, but they cannot come through to p-side due to the interface field. To make the current flow, we have to apply enough positive voltage (0.7 V for silicon) on the p terminal so that it cancels the interface field. Then current can flow. If we apply a positive voltage (reverse bias) to n terminal, we cannot pull the valence of n, as they are filled. Which means valence of p cannot move in to n. We cannot pull conduction of n, as who will take their place, there is no conduction on p. We end up with no current.

Figure 5.2B shows the energy of the valence band and conduction band and fermi level as function of position. Due to interface field, the energy of the bands is shifted upwards on the p side. Figure 5.2C shows the energy of the valence band and conduction band, and fermi level, as function of position under forward bias. Figure 5.2C shows the energy of the valence band and conduction band and fermi level as function of position under reverse bias.

We saw $p - n$ junction conducts when it is forward biased and doesn't conduct in the opposite direction. Therefore a p-n junction is used as a rectifier and is called a diode. It is shown in Fig. 5.3B with arrow pointing toward n region. When diode is forward biased, there is 0.7 V drop across the diode and remaining voltage drop is across resistance. When diode is reverse biased, all drop across the diode and no voltage drop is across resistance. Voltage $V(t)$ across resistance is a rectified version

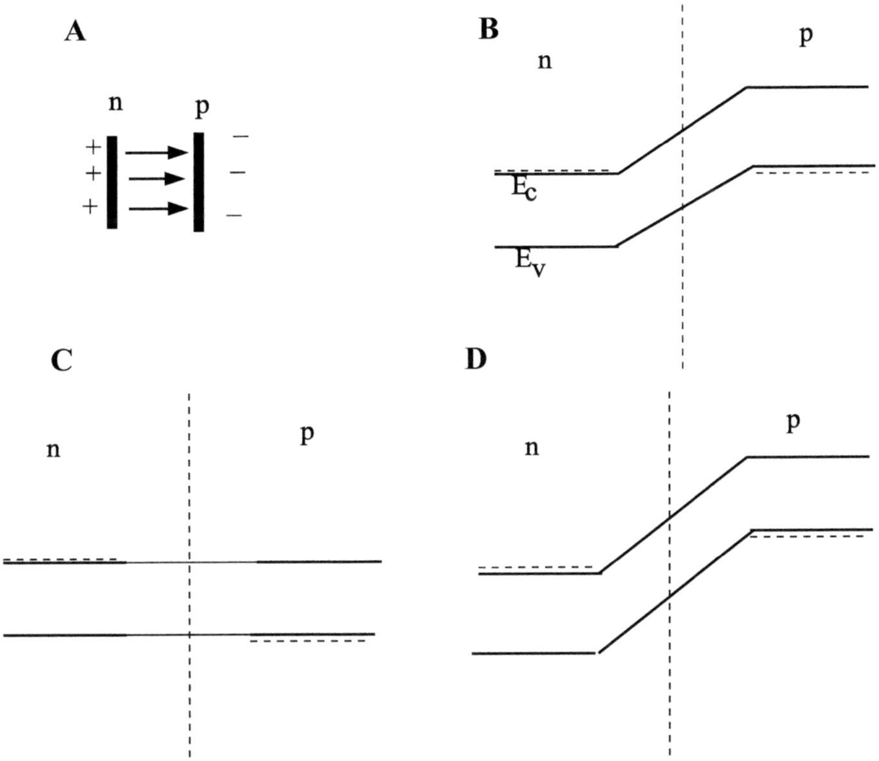

Fig. 5.2 **A** Shows an electric field at the interface. **B** Shows band energy diagram where fermi levels get aligned due to interface field. **C** Shows band energy diagram under forward bias. **D** Shows band energy diagram under reverse bias

of voltage $U(t)$ across source as in Fig. 5.3A. Rectifiers find use in all sorts of circuits including communication circuits.

We saw how to build a rectifier using semiconductor junctions. Its called a diode. Let's now see how to build an amplifier using semiconductor junctions.

5.3 Transistor

Transistor has two junctions as shown in Fig. 5.4A. A p region called base, sandwiched between two n regions (called emitter and collector) as shown in Fig. 5.4A. One pn junction is forward biased (emitter and base) and other reverse biased (base and collector). Forward biased pn junction sucks valence electrons from the p-region and **hurls** conduction electrons from the n-region in the p-region as shown in Fig. 5.4B. These electrons are immediately sucked by the reverse biased junction,

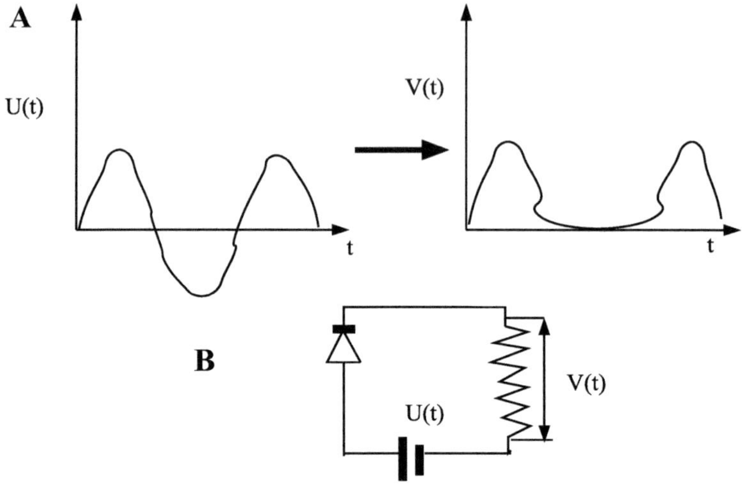

Fig. 5.3 **A** Shows how an voltage is rectified. **B** Shows the circuit to do rectification

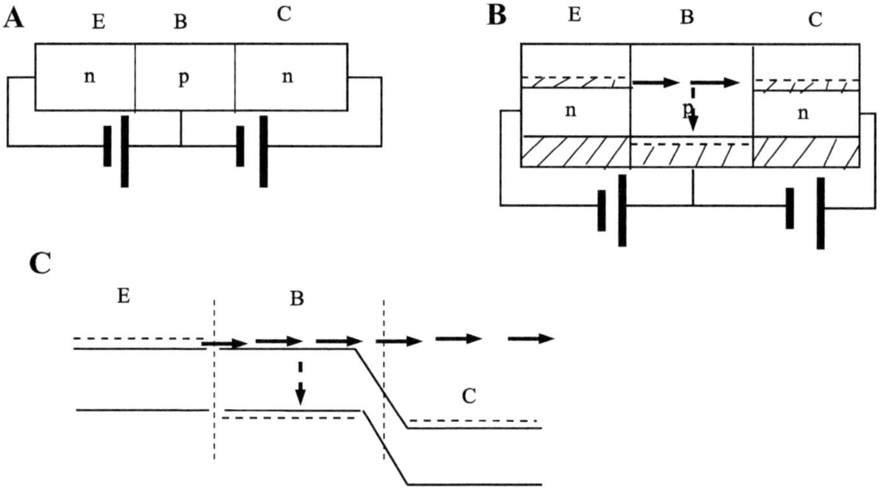

Fig. 5.4 **A** Shows schematic of a transistor. **B** Shows a cartoon of how electrons and bands are in various regions without taking into account interface electric fields. **C** Shows the energy level diagram of a transistor

and on their transit fraction α of them recombine, and contribute to the p junction current, the remaining $1 - \alpha$ fraction just fly by to other n-region as shown in Fig. 5.4B. If I is the base current, $\frac{1-\alpha}{\alpha} I = \beta I$ is the collector current, and $(1 + \beta)I$ the emitter current. α is small, so β is large and is called the current gain of the transistor. Figure 5.4C shows the energy level diagram for the transistor.

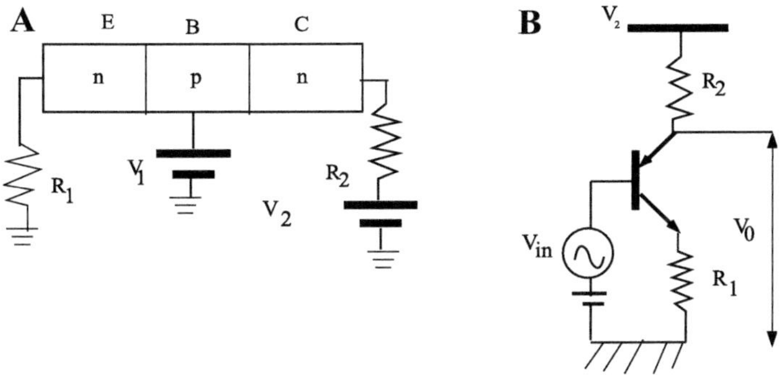

Fig. 5.5 **A** Shows schematic of a transistor. **B** Shows a cartoon of how electrons and bands are in various regions without taking into account interface electric fields. **C** Shows the energy level diagram of a transistor

5.3.1 Current Source and Amplifier

The transistor acts as a current source. Figure 5.5A shows by use of a forward bias and resistance R_1, we establish a current in emitter, simply given by $I_e = \frac{V_1 - 0.7}{R_1}$. This gives a collector current $\frac{\beta}{\beta+1} I_e \sim I_e$ and the potential drop across resistance R_2 as $I_e R_2 = \frac{R_2(V_1 - 0.7)}{R_1}$. Thus the voltage V_1 is amplified to $\frac{R_2}{R_1} V_1$, where $\frac{R_2}{R_1}$ is the amplifier gain. V_1 can have oscillating parts as shown in Fig. 5.5B, which get amplified, the DC part of V_1 is used to bias the base emitter circuit as shown in Fig. 5.5B.

A good amplifier should have a large input impedance as all voltage from a source appears across it. Similarly, a good amplifier should have low output impedance as all amplified voltage appears across the load. Let's compute the input and output impedance of the amplifier in Fig. 5.5B. Changing input voltage by ΔV gives $\Delta I_e = \frac{\Delta V}{R_1}$ and $\Delta I_b = \frac{\Delta V}{\beta R_1}$, then input impedance is $\frac{\Delta V}{\Delta I_b} = \beta R_1$. Similarly, if we measure the voltage across R_2 with a voltmeter with impudence R' then R_2 gets modified to $\frac{R_2 R'}{R_2 + R'}$ and hence the voltage is $\frac{R'}{R_2 + R'} \frac{R_2}{R_1} V_1$, hence the voltage is reduced by a factor $\frac{R'}{R_2 + R'}$, this means output impudence is just R_2.

5.4 Field Effect Transistor

What we have studied is a Bipolar junction transistor. One of its main applications is its use as an amplifier. Now we study another kind of transistor, called Field effect transistor (FET). Its main application is that it acts like a switch. FET is as shown in Fig. 5.8A. There is a p-doped channel, sandwiched between two n-doped regions, called source (S)and drain (D) respectively. If we apply a voltage between source and drain, no current will flow as shown in Fig. 5.8B as p region has only carriers in the valence band, which cannot move to n regions, as valence band is full. For

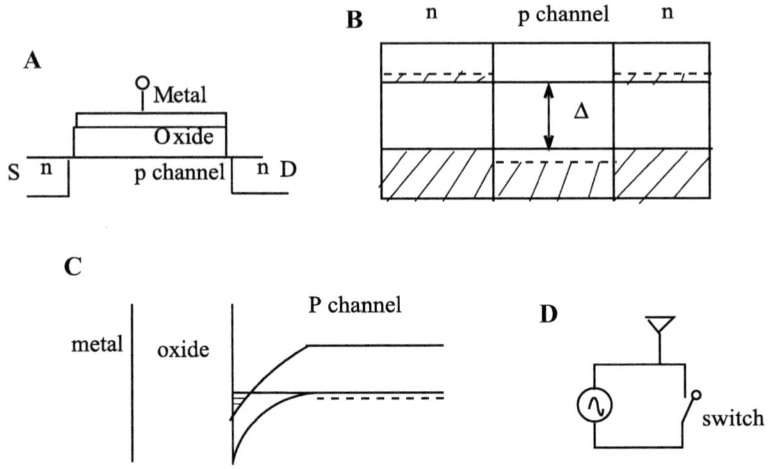

Fig. 5.6 **A** Shows the schematic of a MOSFET. **B** Shows the bands for the source, channel, and drain. **C** Shows the formation of the inversion layer. **D** Shows how FET can be used as a switch in a communication transmit circuit

conduction, we need carriers in the p region, in the conduction band. We arrange this by building a capacitor, a metal plate, and an oxide dielectric and then we put a positive voltage on the metal. The effect is to attract excess charges in the p region, to the oxide-p interface. These excess charges will go to the conduction band, and we will be able to conduct, this is shown in Fig. 5.8C. We say, we have formed an inversion layer at the oxide-channel interface. We say we have turned on our switch, by applying a positive voltage to the metal plate, also called gate. This device is called Metal Oxide Field Effect Transistor (MOSFET). In Fig. 5.8D is shown a cartoon of a communication system, that transmits, when the switch is on (transmit 1) and does not transmit when the switch is off (transmit 0) (Fig. 5.6).

5.4.1 *MESFET, GaAs, and High Mobility Transistors (HEMT)*

We saw how we form an inversion layer, by applying a positive voltage on the gate, pulling electrons. When we remove the voltage, we relax back and the inversion layer disappears. How fast can be switch inversion layer on and off depends on how fast can we accelerate electrons, with an electric field, which is inversely proportional to their effective mass or directly proportional to their mobility μ ($\mu = \frac{e\tau}{m}$, where τ is relaxation time). In high speed communication, we need faster switching and hence we need to have high mobility electrons. Therefore instead of silicon, we use GaAs which has higher mobility. Why low mass and high mobility for GaAs? Recall in

GaAs, the conduction band is the gallium band with Gallium sites indirectly coupled through As. The energy is of the form $\epsilon(k) = \epsilon_0 - 2t\ \cos(ka)$ with $m^{-1} \propto \frac{d^2\epsilon}{dk^2} \propto t\ a^2$. a is large because Ga sites are farther part, and so is transfer integral t which is a measure of how orbitals scatter through As site. Since As is more electronegative than Si, we see larger t and larger mobility.

Therefore we use GaAs, which has larger mobility in the conduction band. The device is simply a n GaAs channel between n based source and drain. Instead of using oxide we simply use a metal gate. If no voltage is applied to the gate then the channel conducts and on application of negative voltage will push electrons away from the interface creating a depletion region and we stop conducting. Due to higher mobility, it is faster to push/pull GaAs as compared to Si so it can be used for faster switching. InAs has even higher mobility because Indium is more delocalized than Galium and has a larger transfer integral.

The mobility of GaAs is still limited because it is n doped (Si replaces Ga). We can think of Si as Ga with extra charge at doping sites which act as scattering impurities and limit mobility. To get better mobility we should use undoped GaAs. But now how will it work as there is no channel to begin with. We can get a channel, if we instead of oxide use n-doped $Al_{1-x}Ga_x$ As. This material has an even larger bandgap than GaAs as Al is more polar. Then electrons will flow from $Al_{1-x}Ga_xAs$ to GaAs creating a inversion layer called 2DEG (2-dimensional electron gas). This makes GaAs channel conducting. By putting a negative voltage on the gate, we repel these electrons and channel disappears. This arrangement has very high mobility and this FET is called High Mobility Transistors (HEMT). It is used for high speed communication applications.

5.5 Optoelectronics

5.5.1 *Light Emitting Diodes and Semiconductor Lasers*

We have already discussed the basic idea of a light emitting diode. We have a direct bandgap material like GaAs. We make a pn junction out of it. Then we put positive voltage on p junction and pull the valence electrons creating a positively charged region which hurls conduction electrons from the n side. These conduction electrons then recombine with holes in the valence band emitting light. Depending on the bandgap ΔE we emit red light in GaAs, green light in GaP, and blue light in GaN (Fig. 5.7).

The emission from LED is not coherent. As electrons fall from conduction to the valence band, they emit light of the same frequency, but the phase of light is not the same, i.e., it is incoherent. Furthermore, the wavevectors of emitted light are in all directions, the emitted light is not directional. However, if we build a cavity that can store this light by reflecting it from its ends. Then the emitted light will induce more emissions. This is called simulated emissions, very much like a two-level system

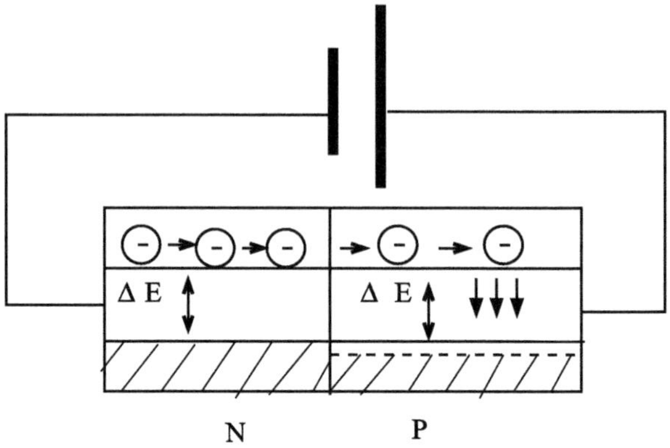

Fig. 5.7 Figure shows the schematic of a LED. A positive voltage on p junction pulls the valence electrons creating positively charged region which hurls conduction electrons from the n side. These conduction electrons then recombine with holes in the valence band emitting light

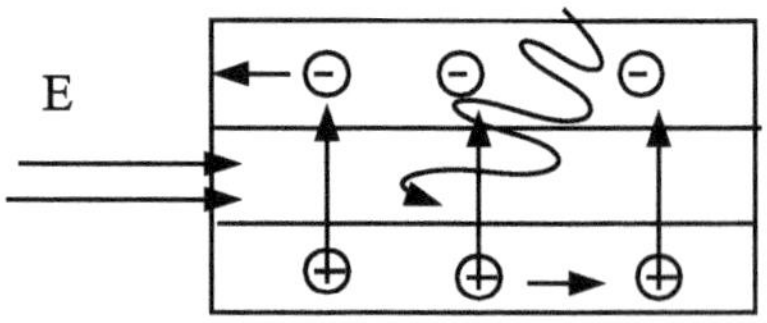

Fig. 5.8 Figure shows the schematic of a solar cell. Light creates electron-hole pairs. The electric field E in depletion region sorts them

driven by an electric field. if λ is the spontaneous emission rate and μ is the rate of simulated emission due to a single photon. Then as soon as a single photon is emitted it induces another emission and does it before another spontaneous emission takes place. Then we have two coherent photons which emit more and so on. If $\mu \gg \lambda$ we will emit many photons before another spontaneous emission and they all will be coherent. Furthermore simulated light has the same wavevectors and hence direction. Thus we have emitted directional and coherent light, this is a laser as in your laser pointer.

5.5.2 Solar Cells

We talked about how in an LED when electrons are hurled in the p region they recombine with the holes and emit light. The opposite of this is that we shine the light of the right wavelength and promote electrons from the valence to the conduction band. Then we can collect these negative and positive carriers and make a battery out of them. For instance in the depletion region of a pn junction, there is an inbuilt electric field. The carriers that are generated in the region will travel to the opposite side due to this field and form a battery. This is a solar cell. It's just a pn junction with light creating electron-hole pairs in the depletion region. Its schematic is shown in the figure.

5.6 Fermi–Dirac Distribution

Consider a half filled band, Lower half is all filled and upper half empty. This is only true at 0 temperature. At finite temperatures we have a probability distribution that even upper orbitals have a finite probability of occupation. This is given by Fermi–Dirac distribution. The probability of orbital with energy ϵ being occupied is given by

$$f(\epsilon) = \frac{1}{1 + \exp\left(\frac{\epsilon - \mu}{kT}\right)},$$

where μ is a parameter called chemical potential. At $T = 0$, $\mu = \epsilon_F$ the fermi energy, else it is slightly above it. Then you can see at $T = 0$ nothing above ϵ_F is filled. Going by this formula even high unfilled bands have finite occupancy. If we consider a semiconductor like silicon, where the valence band is full and the conduction band is empty, at finite temperatures we find finite occupancy in the conduction band which increases at high temperatures so that at temp like 1000 K, silicon is a conductor because conduction band has sizeable carriers and at low temperatures, it is an insulator. This is shown in Fig. 5.9. We can calculate the concentration of carriers in conduction band n. if we take bottom of conduction band to be at 0 energy then this is simply

$$\int_0^\infty N(\epsilon) f(\epsilon), \tag{5.1}$$

where $N(\epsilon)$ is density of states at ϵ. Recall $\epsilon = \frac{\hbar^2 k^2}{2m}$, then $\frac{\Delta\epsilon}{\epsilon} = 2\frac{\Delta k}{k}$. number of states in band of width Δk is $4\pi k^2 \Delta k \frac{V}{(2\pi)^3} = V \frac{(2m)^{\frac{3}{2}}}{\hbar^3 (2\pi)^2} \sqrt{\epsilon} \Delta\epsilon$.

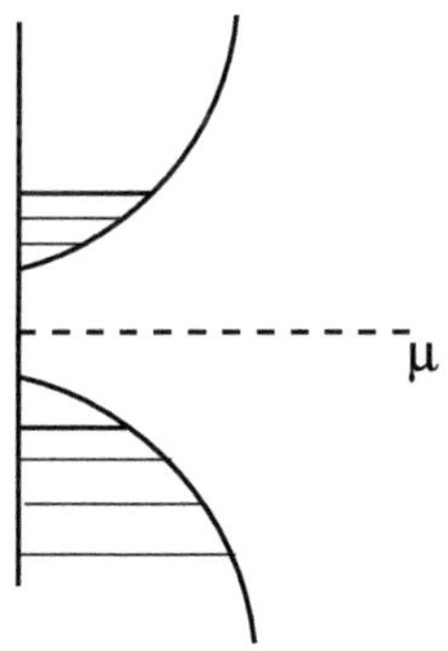

Fig. 5.9 Figure shows electrons and holes in conduction and valence band respectively at finite temperatures

Then

$$n = \frac{(2m)^{\frac{3}{2}}}{\hbar^3(2\pi)^2} \int_{E_c}^{\infty} \sqrt{\epsilon} \exp\left(-\frac{(\epsilon-\mu)}{kT}\right) = \sqrt{2}\frac{m^{\frac{3}{2}}}{\hbar^3\pi^2} \exp\left(-\frac{E_C-\mu}{kT}\right) \int_0^{\infty} x^2 \exp\left(-\frac{x^2}{kT}\right) = N_C \exp\left(-\frac{E_C-\mu}{kT}\right) \tag{5.2}$$

N_c is around $10^{19}/\text{cm}^3$ for Si at 300 K.

Similarly we can derive a formula for hole concentration p and is given by

$$p = N_V \exp\left(\frac{E_v - \mu}{kT}\right). \tag{5.3}$$

Then the product using $n = p = n_i$

$$n_i^2 = np = N_C N_V \exp\left(-\frac{E_g}{kT}\right) \tag{5.4}$$

n_i is called intrinsic carrier concentration and depends on bandgap E_g. using $N_c = N_V \sim 10^{19}/\text{cm}^3$, and $E_g = 1.1$ eV, $n_i = 10^{10}$ at room temperature.

If we dope say with donor concentration N_D, then $\mu \to \mu'$ and we can show we still have $np = n_i^2$. With $n = N_D$, we can find p. Furthermore

$$E_c - \mu' = kT\ ln(N_C/N_D). \tag{5.5}$$

If instead we dope with acceptors with concentration N_A.

$$\mu' - E_v = kT\ ln(N_V/N_A). \tag{5.6}$$

Problems

1. We considered tight binding in the mono-atomic chain. We now consider diatomic chain

$$A - B - A - B - \cdots - A - B$$

   where onsite energy at site A is ϵ_A and site B is ϵ_B and transfer element is $-t$. Find the dispersion relation and sketch it for this chain.
2. A silicon ingot is doped with 10^{16} arsenic atoms/cm^3. Find the carrier concentrations and the Fermi level at room temperature (300 K).
3. Calculate the inbuilt potential for a silicon pn junction with $N_A = 10^{18}/\text{cm}^3$ and $N_D = 10^{15}/\text{cm}^3$ at 300 K.

4. In Fig. 5.5B of the book assumes DC voltage on the base of 1 V (no ac voltage). Find the collector current if the $\beta = 100$ and $R_1 = 100\ \Omega$.
5. In above what should be R_2 for an amplifier gain to be 100.
6. Consider metal calcium or magnesium with two electrons it the outer shell. This says that the conduction band will be full. Why is it a metal then?

References

1. S.M. Sze, *Semiconductor Devices: Physics and Technology* (Wiley, 1985)
2. P. Horowitz, W. Hill, *Art of Electronics* (Cambridge University Press, 1989)

Chapter 6
Superconductivity

There is a very interesting phenomenon that takes place in solid state physics, when certain metals are cooled below the critical temperature of order of few kelvin. The resistance of these metals completely disappears, and they become superconducting. This phenomenon whereby many materials exhibit complete loss of electrical resistance, when cooled below a characteristic critical temperature [1, 2] is called superconductivity. It was discovered in mercury, by Dutch physicist Onnes in 1911. For decades, a fundamental understanding of this phenomenon eluded the many scientists who were working in the field. Then, in the 1950s and 1960s, a remarkably complete and satisfactory theoretical picture of the classic superconductors emerged in terms of the Bardeen Cooper Schrieffer (BCS) theory [3].

In this chapter, we spell out the main ideas of the BCS theory. BCS theory tells us how to use phonon mediated interaction to bind electrons together, so that we have a big molecule, which we call the BCS ground state or the BCS molecule. At low temperatures, phonons do not have the energy to break the bonds, in the molecule, hence electrons in the molecule do not scatter phonons. So let's see how BCS binds these electrons into something big.

Figure 6.1 shows how it is when we accelerate electrons with an electric field say in $-x$ direction. The whole Fermi sphere displaces to the right by a small amount. There is net momentum in the x direction, and this constitutes the current. How does the current stop? The electrons on the right, shown as black dots in Fig. 6.1 are scattered to the left as shown. This scattering is due to the absorption of phonons and annuls the forward x-momentum of electrons. The current stops. Figure 6.2A shows that in superconductivity, an antipodal electron pair forms a bond. Resonant absorption of phonon as in Fig. 6.2B breaks this bond. At low temperatures, phonons do not have enough energy to break this bond, i.e., they are *off-resonant*. Then scattering of electrons does not happen, and the current does not stop, and we have superconductivity. In the remaining chapter, we study this bond between antipodal electrons.

N. Khaneja, *Electron Waves in Solids*, Springer Series in Solid-State Sciences 203,
https://doi.org/10.1007/978-3-031-67260-6_6

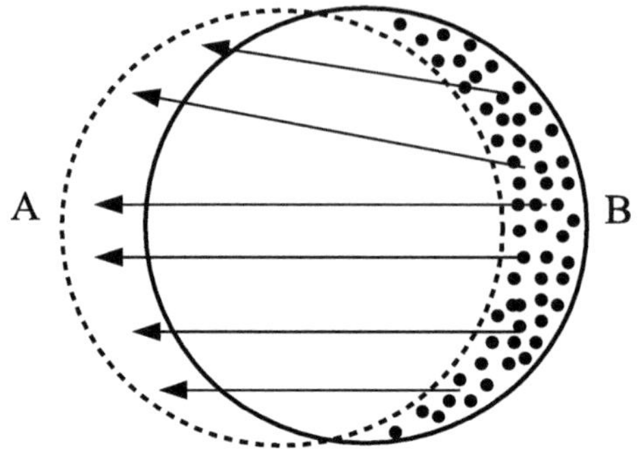

Fig. 6.1 Figure shows how electrons are scattered by absorption of phonons

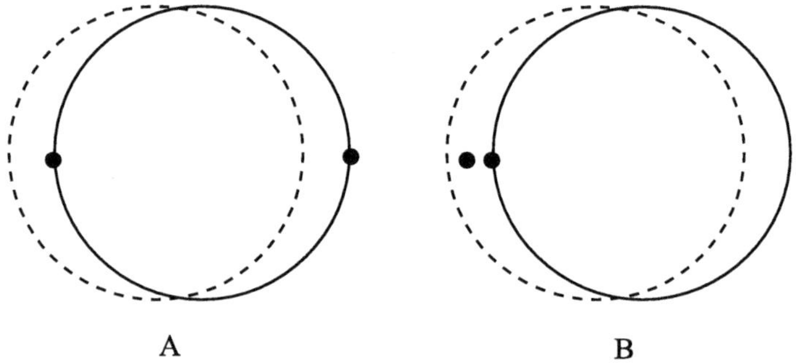

Fig. 6.2 Figure shows how resonant absorption of phonon breaks a pair of antipodal electrons as in **A** to as in **B**, something superconductivity resists

6.1 Cooper Pairs and Binding

Let us take two electrons, both at the Fermi surface, one with momentum k_1 and the other $-k_1$. Let's see how they interact with phonons. Electron k_1 pulls/plucks on the lattice due to Coulomb attraction, and in the process creates (emits) a phonon and thereby recoils to new momentum k_2. The resulting lattice vibration is sensed by electron $-k_1$ which absorbs this oscillation and is thrown back to momentum $-k_2$. The total momentum is conserved in the process. This is depicted in Fig. 6.3A. The corresponding Feynman diagram for this process is shown in Fig. 6.3B. The above process, where two electrons interact, with the exchange of phonon can be represented as a three-level atomic system. Level 1 is initial state of the electrons $k_1, -k_1$ and level 3 is the final state of the electrons $k_2, -k_2$ and the level 2 is the intermediate state $k_2, -k_1$. There is transition with strength $\Omega = \hbar d$ between level 1 and 2, involving emission of a phonon and a transition with strength Ω between level 2 and 3 involving absorption of a phonon. Now let's pause, and remember what is Ω. It comes from electron–phonon interaction as discussed in Chap. 3. The electron–phonon coupling Hamiltonian (*Fröhlich* Hamiltonian) is of form

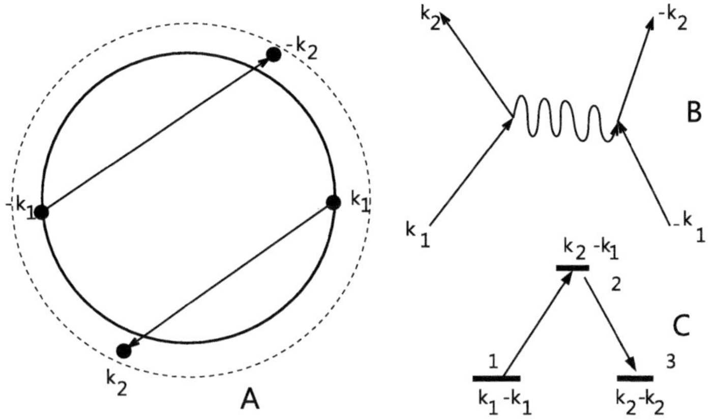

Fig. 6.3 **A** depicts the Fermi sphere and how electron pair $k_1, -k_1$ at Fermi sphere scatters to $k_2, -k_2$ at the Fermi sphere. **B** shows how this is mediated by the exchange of a phonon in a Feynman diagram. **C** shows a three-level system that captures the various transitions involved in this process

$$\underbrace{\frac{c}{\sqrt{n^3}}}_{\Omega} i(b \exp(ikx) - b^{\dagger} \exp(-ikx)), \tag{6.1}$$

where $b, b^{\dagger}$ are annihilation and creation operators for phonon. We estimated $c \sim 1$ V.

Now let E_1, E_2, E_3 be the energy of the three levels as in Fig. 6.3C. $E_1 = 2\epsilon_1 = 2\hbar\omega_1$, $E_3 = 2\epsilon_2 = 2\hbar\omega_2$ and $E_2 = \epsilon_1 + \epsilon_2 + \underbrace{\hbar\omega_d}_{\epsilon_d}$. The state of the three-level system evolves according to the Hamiltonian

$$H = \underbrace{\begin{bmatrix} E_1 & 0 & 0 \\ 0 & E_2 & 0 \\ 0 & 0 & E_3 \end{bmatrix}}_{H_0} + \Omega \underbrace{\begin{bmatrix} 0 & 1 & 0 \\ 1 & 0 & 1 \\ 0 & 1 & 0 \end{bmatrix}}_{H_1}. \tag{6.2}$$

Let e_1, e_2, e_3 be coordinate vectors which are eigenvectors of H_0 with eigenvalues E_1, E_2, E_3 respectively. H_1 is a perturbation which to first order perturbs these eigenvectors to

$$e_1 \rightarrow e_1 + \alpha e_2, \tag{6.3}$$
$$e_2 \rightarrow e_2 - \alpha e_1 - \beta e_3, \tag{6.4}$$
$$e_3 \rightarrow e_3 + \beta e_2, \tag{6.5}$$

where $\alpha = \frac{\Omega}{E_1 - E_2}$ and $\beta = \frac{\Omega}{E_3 - E_2}$. We make these perturbed eigenvectors orthonormal to second order as

$$e_1 \to e_1(1 - \frac{\alpha^2}{2}) + \alpha e_2 - \frac{\alpha\beta}{2} e_3, \tag{6.6}$$

$$e_2 \to e_2(1 - \frac{\alpha^2 + \beta^2}{2}) - \alpha e_1 - \beta e_3, \tag{6.7}$$

$$e_3 \to e_3(1 - \frac{\beta^2}{2}) + \beta e_2 - \frac{\alpha\beta}{2} e_1, \tag{6.8}$$

and now we write H in these new basis as

$$H' = \begin{bmatrix} E_1 + \alpha\Omega & 0 & \frac{\alpha+\beta}{2}\Omega \\ 0 & E_2 - (\alpha+\beta)\Omega & 0 \\ \frac{\alpha+\beta}{2}\Omega & 0 & E_3 + \beta\Omega \end{bmatrix}. \tag{6.9}$$

We find there is second order coupling between state 1 and 3 given by

$$H'_{13} = \frac{\alpha+\beta}{2}\Omega = \frac{1}{2}\left(\frac{\Omega^2}{E_1 - E_2} + \frac{\Omega^2}{E_3 - E_2}\right) = \hbar \frac{d^2 \omega_d}{\underbrace{(\omega_1 - \omega_2)^2}_{\Delta\omega} - (\omega_d)^2}. \tag{6.10}$$

When $\epsilon_1 = \epsilon_2$, we get

$$H'_{13} = -\hbar \frac{d^2}{\omega_d}. \tag{6.11}$$

This effective coupling gives us the attractive potential responsible for superconductivity [3]. We say the electron pair $k_1, -k_1$ scatters to $k_2, -k_2$ at rate $-\hbar\frac{d^2}{\omega_d}$. The scattering rate is in fact $V = -\hbar\frac{4d^2}{\omega_d}$ as k_1 can emit to k_2 or $-k_2$. Similarly $-k_1$ can emit to k_2 or $-k_2$, making it total of four processes that can scatter $k_1, -k_1$ to $k_2, -k_2$.

How does all this help? Suppose $|k_1, -k_1\rangle$ and $|k_2, -k_2\rangle$ are only two states around of same energy 2ϵ. Then a state like

$$\phi = \frac{|k_1, -k_1\rangle + |k_2, -k_2\rangle}{\sqrt{2}} \tag{6.12}$$

has energy $2\epsilon + V$. That is lower energy than the individual states in the superposition. V is the binding energy.

We said there are only two states, $|k_1, -k_1\rangle$ and $|k_2, -k_2\rangle$. In general, we have for $i = 1, \ldots, \mathcal{N}$, $|k_i, -k_i\rangle$ states on Fermi sphere as shown in Fig. 6.4A and if we form the state

$$\phi = \frac{1}{\sqrt{\mathcal{N}}} \sum_i |k_i, -k_i\rangle, \tag{6.13}$$

it has energy $2E_F + (\mathcal{N} - 1)V$ where E_F is fermi energy.

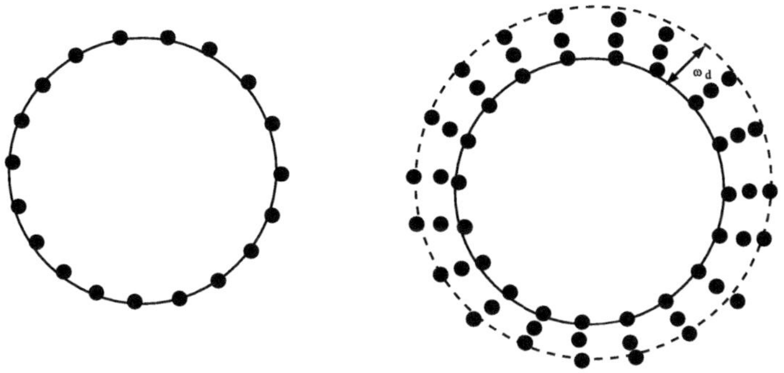

Fig. 6.4 **A** shows electron pairs on the fermi surface. **B** shows electron pairs in an annulus around fermi surface

In (6.10), as long $\Delta\omega < \omega_d$, in BCS theory, we approximate $V \sim -\frac{4\Omega^2}{\epsilon_d}$. Therefore, if we take an annulus in Fig. 6.4B to be of width ω_d, we get total number of states in the annulus $\mathcal{N}$ to be $\frac{\mathcal{N}}{n^3} \sim \frac{\omega_d}{\omega_F}$, where n^3 is total number of k points in the fermi sphere and $E_F = \hbar\omega_F$ is fermi energy. This gives a binding energy $\Delta_b =\sim \frac{c^2}{E_F}$. With the fermi energy $E_F \sim 5 - 10$ eV, the binding energy $\sim$ meV. Thus we have shown how phonon mediated interaction helps us bind an electron pair with energy $\sim$ meV. This paired electron state is called a Cooper pair.

6.2 Cooper Problem

Consider a electron pair $|k, -k\rangle$ at the Fermi surface with energy $2(E_F + \xi_k)$ that scatters to states $|l, -l\rangle$ in an annulus of width $\hbar\omega_d$ at fermi surface with rate V. Then let's compute the minimum energy and corresponding eigenvector. Writing the eigenvector as

$$\phi = \sum_k x_k |k, -k\rangle. \tag{6.14}$$

The eigenvalue equation reads

$$2\xi_k x_k - \sum_l V x_l = E x_k, \tag{6.15}$$

or

$$x_k = \frac{V \sum_l x_l}{2\xi_k - E}, \tag{6.16}$$

summing over k, with D as density of states at Fermi surface, we get

$$\frac{1}{DV} = \int_0^{\hbar\omega_d} \frac{1}{2\xi - E} d\xi. \tag{6.17}$$

This gives for $DV \ll 1$, we get

$$E = -2\hbar\omega_d \exp(-\frac{1}{DV}). \tag{6.18}$$

Thus we get that energy is reduced below the Fermi energy of $2E_F$.

6.3 Dummy Ground State

In Cooper's problem, we talked about a superposition state of two electrons, that has smaller energy than the Fermi energy of $2E_F$. This superposition was formed from two antipodal electrons, on Fermi sphere and antipodal states in an annulus around Fermi surface. But we do not have just two electrons. What should be the state of all the electrons near the Fermi surface so that there is enough room to scatter to empty states for all the electron pairs present, so as to reduce the total energy. For this we propose a simple solution, we call the dummy ground sate. It highlights the essence of real solution to the problem studied in the next section. Imagine an annulus of energy width $\hbar\omega_d$ around the Fermi sphere as shown in Fig. 6.5. We have say $2N$ possible antipodal pairs in this annulus. Let's say half of them N are occupied and remaining empty (Clearly the occupied ones have come for filled states in annulus of width $\frac{\hbar\omega_d}{2}$

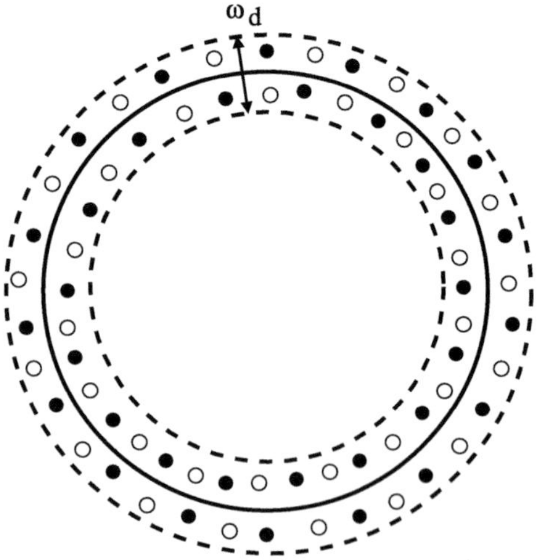

Fig. 6.5 Figure shows the dummy ground state

below the Fermi surface in normal conductor). Let's denote the occupied states as $\phi_{i_1}\phi_{i_2}\ldots\phi_{i_N}$. Not consider a massive superposition of all such half filled states.

$$\Phi = \frac{1}{Z}\sum_{i_1,\ldots,i_N} \phi_{i_1}\phi_{i_2}\ldots\phi_{i_N}, \tag{6.19}$$

where Z is appropriate normalization to get norm 1 for Φ.

There are N empty pairs, ϕ_{i_1} can scatter to. Similarly ϕ_{i_2}, and so on. Then there are N^2 states that can scatter to the state $\phi_{i_1}\phi_{i_2}\ldots\phi_{i_N}$. This is true of all the terms in Φ, and hence Φ is an eigenvector of the pair wise scattering Hamiltonian with eigenvalue $-N^2V$. Thus potential energy of the state Φ in (6.19) is $-N^2V$. Each term in (6.19) can be reflected around the Fermi surface. So that if ϕ_{i_1} has kinetic energy $2E_F - 2\xi$, then the reflected state has energy $2E_F + 2\xi$. This ensures that the total kinetic energy of the state Φ in (6.19) is simply $2NE_F$. Then the total energy is

$$E_\Phi = 2NE_F - N^2 V. \tag{6.20}$$

What is the energy of the normal state, in which all pairs are in the annulus of width $\frac{\hbar\omega_d}{2}$ below the Fermi surface. It is simply $2NE_F - N\frac{\hbar\omega_d}{2}$. Then the difference is

$$E_\Phi - E_N = -N^2 V + N\frac{\hbar\omega_d}{2}. \tag{6.21}$$

But note $N = D\frac{\hbar\omega_d}{2}$. This means

$$E_\Phi - E_N = \frac{\hbar\omega_d\, D}{4}(-DV + 1). \tag{6.22}$$

If $DV > 1$, which we call a strong coupling regime, we reduce the total energy by forming the dummy state and this state has lower energy than the normal state. However if $DV < 1$, which is what is called the *weak coupling limit*, we increase then energy over the normal state. However, there is another solution to the arrangement of pairs, which even gives lower energy than normal energy in a weak coupling limit. It is called the BCS ground state that we describe in the next section. Before we proceed we mention two interesting facts.

Elementary excitations: Suppose in our dummy state, we introduce an extra electron in the plane wave state $|k\rangle$, at Fermi surface. Then that renders the pair ϕ_k out of service as we cannot scatter in and out of this pair. Then if we compute E_Φ, we can only scatter to remaining N_1 pairs, hence total energy is

$$E_\Phi = (2N + 1)E_F - N(N - 1)V. \tag{6.23}$$

This means that the introduction of an extra electron at the Fermi surface has raised the energy not just by E_F but $E_F + \Delta$ where $\Delta = NV$.

Broken pair: Suppose we break an antipodal state $|k, -k\rangle$ into plain wave state $|k\rangle$ and $|l\rangle$ at Fermi surface. This renders two pairs out of service. Of the remaining $N-1$ electron pairs, we form a state Φ' where each pair can scatter to $N-1$ pairs. Thus the total energy is

$$E_\Phi = 2NE_F - (N-1)^2\, V. \tag{6.24}$$

Thus the total energy rises by 2Δ. Now we study the weak coupling limit $DV < 1$ and the BCS ground state.

6.4 BCS Ground State

Once again, imagine an annulus of energy width $\hbar\omega_d$ around the Fermi sphere, as shown in Fig. 6.5. We have say $2N$ possible antipodal pairs in this annulus. In the last section, we said half of them, N, are occupied. Now let's relax this hard constraint and say pair ϕ_k is occupied with probability v_k^2 and empty with probability u_k^2 and postulate the state

$$\Phi_0 = \prod_k (u_k + v_k\phi_k) = \sum c_{i_1,\ldots,i_n}\, \phi_{i_1}\ldots\phi_{i_n}, \tag{6.25}$$

where k runs over $2N$ possible antipodal pairs in this annulus. Let's find u_k, v_k so that energy E_0 of Φ_0 is minimized.

$$E_0 = 2\sum_k \xi_k v_k^2 - V\sum_{k,l} u_k v_k u_l v_l. \tag{6.26}$$

$$\Phi_0 = \prod_k (\cos\theta_k + \sin\theta_k\phi_k). \tag{6.27}$$

$$E_0 = 2\sum_k \xi_k \cos^2\theta_k - \frac{1}{4}V\sum_{k,l} \sin 2\theta_k \sin 2\theta_l. \tag{6.28}$$

Minimize E_0,

$$0 = \frac{\partial E}{\partial\theta_k} = 2\xi_k \sin 2\theta_k + V\cos 2\theta_k \sum_l \sin 2\theta_l. \tag{6.29}$$

Let

$$\Delta_l = \frac{V}{2}\sum_l \sin 2\theta_l, \tag{6.30}$$

which we take as same for all l, i.e., $\Delta_l = \Delta$.

Let $E_k = \sqrt{\Delta^2 + \xi_k^2}$, then

$$\tan 2\theta_k = -\frac{V}{2\xi_k} \sum_l \sin 2\theta_l. \tag{6.31}$$

This gives

$$\tan 2\theta_k = -\frac{\Delta}{\xi_k}. \tag{6.32}$$

$$2u_k v_k = \sin 2\theta_k = \frac{\Delta}{E_k}. \tag{6.33}$$

$$v_k^2 - u_k^2 = \cos 2\theta_k = -\frac{\xi_k}{E_k}. \tag{6.34}$$

Then we get

$$\sum_k \frac{1}{E_k} = \frac{1}{V}. \tag{6.35}$$

We can sum the left side. Let D denote density of states then

$$\frac{1}{DV} = \int_0^{\frac{\hbar\omega_d}{2}} \frac{d\xi}{\sqrt{\xi^2 + \Delta^2}} = \frac{1}{\Delta} \sinh^{-1}(\frac{\hbar\omega_d}{2}). \tag{6.36}$$

$$\Delta = \frac{\hbar\omega_d}{2\sinh(\frac{1}{DV})} \sim \hbar\omega_d \exp(-\frac{1}{DV}). \tag{6.37}$$

6.4.1 Energy of the Ground State

Having determined the coefficients u_k, v_k, we can ask what is the energy of the ground state. By simply using definition of Δ in (6.30),

$$E_0 = 2 \sum_{|k|<k_F} \xi_k v_k^2 + 2 \sum_{|k|>k_F} \xi_k v_k^2 - \frac{\Delta^2}{V}. \tag{6.38}$$

If we subtract from it the energy of the normal state E_n where the fermi sphere is all filled we get

$$E_n = 2 \sum_{|k|<k_F} \xi_k. \tag{6.39}$$

We get by substituting for $v_k^2 = \frac{1}{2}(1 - \frac{\xi_k}{E_K})$,

$$E_0 - E_n = 2 \sum_{|k|>k_F} (\xi_k - \frac{\xi_k^2}{E_k}) - \frac{\Delta^2}{V}. \tag{6.40}$$

$$E_0 - E_n = 2 \sum_{|k|>k_F} (\xi_k - E_k) + \frac{\Delta^2}{V}. \tag{6.41}$$

Evaluating

$$2 \sum_{|k|>k_F} (\xi_k - E_k) = 2D \int_0^{\frac{\hbar\omega_d}{2}} \xi - \sqrt{\xi^2 + \Delta^2} \tag{6.42}$$

$$= 2D \int_0^{\Delta} \xi - \sqrt{\xi^2 + \Delta^2} + \int_{\Delta}^{\frac{\hbar\omega_d}{2}} \xi - \sqrt{\xi^2 + \Delta^2} \tag{6.43}$$

$$\sim -\frac{D\Delta^2}{2} - \frac{\Delta^2}{V}. \tag{6.44}$$

Then

$$E - E_n \sim -\frac{D\Delta^2}{2}. \tag{6.45}$$

6.4.2 Energy of Elementary Excitation

Suppose we add an extra electron in the state $\psi_m = |m\rangle$, then it forbids Cooper pair ϕ_m. The resulting state is

$$\Phi_e = \prod_{k \neq m} (u_k + v_k \phi_k) \psi_m. \tag{6.46}$$

Let's calculate the energy of Φ_e. We do not have state ϕ_m to scatter to Cooper pairs which raise the energy by

$$\frac{V}{2} \sin\theta_m \sum \sin\theta_l = \sin 2\theta_m \Delta.$$

The change of energy then is

$$E_e - E_0 = E_F + \xi_m - 2v_m^2 \xi_m + \sin 2\theta_m \Delta, \tag{6.47}$$

$$= E_F + E_k. \tag{6.48}$$

Thus introduction of an extra electron in the superconducting state, raises the energy by a minimum of $E_F + \Delta$. If we break a pair which amounts to removing states ϕ_m and ϕ_n to scatter to. This means we raise the energy by $2E_k$ the minimum of which is 2Δ.

6.4.3 Finite Temperatures

At finite temperature, there is the probability that we have elementary excitation of the state $|k\rangle$ which happens with probability ($\beta^{-1} = kT$)

$$f(E_k) = \frac{1}{1 + \exp(\beta E_k).} \tag{6.49}$$

Similarly elementary excitation of state $|-k\rangle$ which happens with probability $f(E_k)$ and the total probability that there is excitation in k pair is $2f(E_k)$ and no excitation is $1 - 2f(E_K)$. Then the energy of the ground state in the presence of these elementary excitations is

$$E_0 = 2f(E_k)\xi + 2\xi_k v_k^2(1 - 2f(E_k)) - V\sum_{kl} u_k v_k u_l v_l(1 - 2f(E_k))(1 - 2f(E_l)), \tag{6.50}$$

again denoting $u_k = \cos\theta_k$ and $v_k = \sin\theta_k$, we find that variation gives

$$\tan 2\theta_k = -\frac{V}{2\xi_k}\sum_l \sin 2\theta_l(1 - 2f(E_l)). \tag{6.51}$$

As before define

$$\Delta = \frac{V}{2}\sum_l \sin 2\theta_l(1 - 2f(E_l)), \tag{6.52}$$

$$\tan 2\theta_k = -\frac{\Delta}{\xi_k}, \tag{6.53}$$

$$\sin 2\theta_k = \frac{\Delta}{E_k}, \tag{6.54}$$

which gives

$$\sum_k \frac{\tanh(\beta E_k/2)}{E_k} = \frac{2}{V}, \tag{6.55}$$

Fig. 6.6 Plot of $\Delta(T)$ as function of T

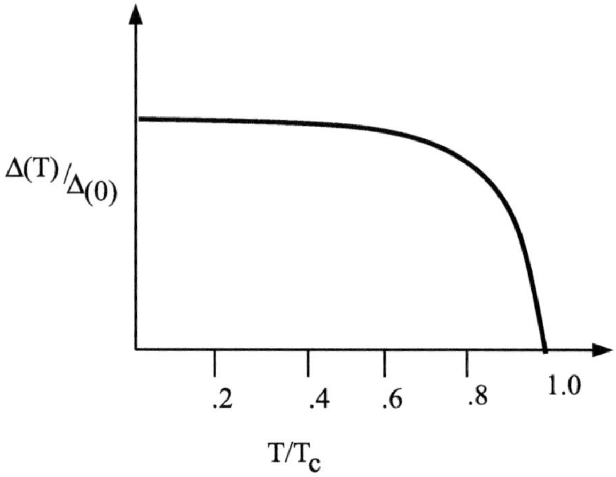

which reads

$$\int_0^{\frac{\hbar\omega_d}{2}} \frac{\tanh(\beta\sqrt{\xi^2+\Delta^2}/2)}{\sqrt{\xi^2+\Delta^2}} = \frac{1}{DV}. \tag{6.56}$$

At T_c, $\Delta \to 0$, which gives

$$\int_0^{\frac{\hbar\omega_d}{2}} \frac{\tanh(\beta_c\xi/2)}{\xi} = \frac{1}{DV}, \tag{6.57}$$

which is evaluated to give

$$\beta_c^{-1} = kT_c = 1.13\frac{\hbar\omega_d}{2}\exp(-\frac{1}{DV}), \tag{6.58}$$

which gives

$$\frac{kT_c}{\Delta(0)} = 1.13/2, \tag{6.59}$$

or $\Delta(0) = 1.76kT_c$. From (6.56), we can plot $\Delta(T)$ as shown in Fig. 6.6.

6.5 Meisner Effect

When a superconductor placed in a magnetic field is cooled below its critical T_c, we find it expels all magnetic fields from its inside. It doesn't like a magnetic field in its interior. This is shown in Fig. 6.7

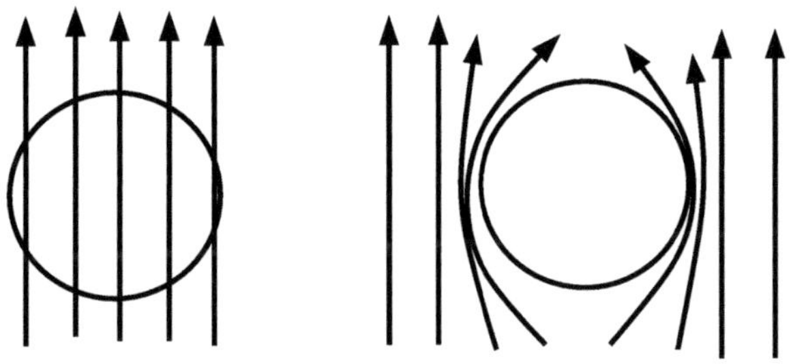

Fig. 6.7 Figure depicts the Meisner effect whereby the magnetic field inside a superconductor is expelled when we cool it below its superconducting temperature T_c

The German physicists, Walther Meissner and Robert Ochsenfeld discovered this phenomenon in 1933, by measuring the magnetic field distribution outside superconducting tin and lead samples. The samples, in the presence of an applied magnetic field, were cooled below their superconducting transition temperature, whereupon the samples canceled nearly all interior magnetic fields. A superconductor, with little or no magnetic field within it is said to be in the Meissner state. The Meissner state breaks down when the applied magnetic field is too large. Superconductors can be divided into two classes according to how this breakdown occurs. In Type-I superconductor, if the magnetic field is above a certain threshold H_c, no expulsion takes place. In type-II superconductors, raising the applied field past a critical value H_{c1} leads to a mixed state (also known as the vortex state) in which an increasing amount of magnetic flux penetrates the material, but there remains no resistance to the electric current as long as the current is not too large. At a second critical field strength H_{c2}, no magnetic field expulsion takes place. How can we explain the Meisner effect?

Let's say we switch on a magnetic field inside a superconductor, then the act induces an electric field according to Maxwell's equations

$$\nabla \times E = -\frac{\partial B}{\partial t}. \tag{6.60}$$

This induced electric field accelerates superconducting electrons as

$$m\frac{dv}{dt} = -eE. \tag{6.61}$$

If we write the current density as $J = -nev$, we get

$$\Lambda\frac{dJ}{dt} = E, \tag{6.62}$$

where $\Lambda = \frac{m}{ne^2}$. Then from (6.60), we get

$$\Lambda\nabla \times J = -B. \tag{6.63}$$

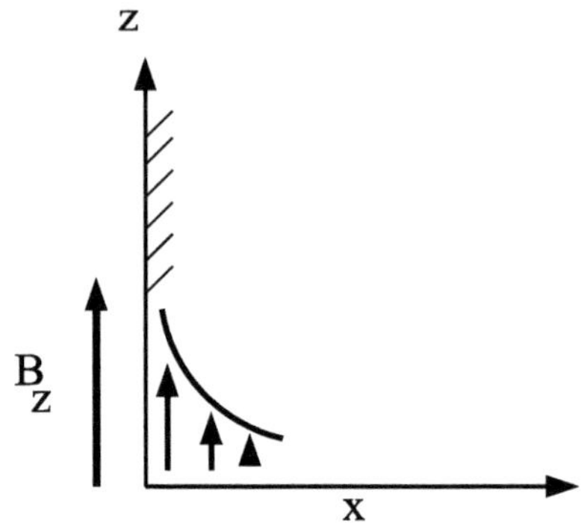

Fig. 6.8 Figure depicts the prefect diamagnetism of a superconductor where an applied magnetic field in z direction decays exponentially inside the superconductor as we move along x direction

We can write $B = B_0 + B_0'$ where B_0 is constant external field and B_0' is produced by electrons. Then we have

$$\nabla \times B_0' = \mu_0 J, \tag{6.64}$$

which gives

$$\nabla \times \nabla \times B_0' = -\frac{\mu_0 B}{\Lambda}, \tag{6.65}$$

or

$$\nabla^2 B_0' = \frac{\mu_0 B}{\Lambda} \tag{6.66}$$

or

$$\nabla^2 B = \frac{\mu_0 B}{\Lambda} = \frac{B}{\lambda_l^2}, \tag{6.67}$$

which gives exponential decay in B inside the conductor. For example see Fig. 6.8 with superconductor as $x > 0$. For a field $B = B_z \hat{z}$, the field decays as $B_z \exp(-\frac{x}{\lambda_l})$. λ_l is called the London penetration depth.

6.6 Giaver Tunneling

When we bring two metals in proximity, separated by a thin insulating barrier, and apply a tiny voltage then the current will flow in the circuit. There is a thin insulating barrier, but electrons will tunnel through the barrier. Now what will happen if one of these metals is replaced by a superconductor. These are the set of experiments carried out by Norwegian-American physicist Ivar Giaver who shared the Nobel Prize in Physics in 1973 with Leo Esaki and Brian Josephson "for their discoveries regarding tunneling phenomena in solids". What he found was, that if one of the metals is a superconductor, the electron cannot just come in, as there is an energy barrier of Δ, the superconducting gap. Your applied voltage has to be at least as big as Δ for tunneling to happen. This is depicted in Fig. 6.9. Let's see why this is the case.

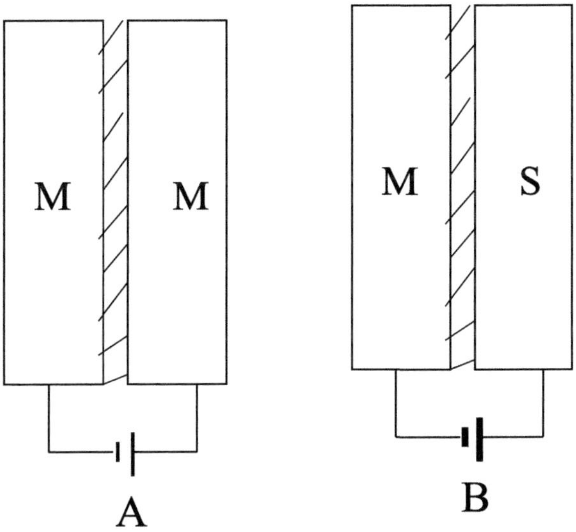

Fig. 6.9 **A** shows how a tiny voltage between two metals separated by an insulating barrier generates a current that goes through the insulating barrier through tunneling. **B** shows if one of the metals is a superconductor, then the applied voltage has to be at least as big as the superconducting gap

Recall in our discussion of the superconducting state, if we have N superconducting electrons, with E_s energy, and we break a pair. Then as discussed in Sect. 6.4.2, we supply a minimum energy of 2Δ. When the broken pair is at Fermi energy the energy of the system is $(N-2)E_s + 2E_F$ and we have

$$NE_s + 2\Delta = (N-2)E_s + 2E_F, \tag{6.68}$$

giving

$$E_s = E_F - \Delta. \tag{6.69}$$

Thus superconducting electrons are Δ energy below the Fermi energy and an incoming electron has an minimum energy $E_F + \Delta$. Therefore we need to apply a voltage as big as $\frac{\Delta}{e}$ for this electron to tunnel in. All this is depicted in Fig. 6.10B.

6.7 Superconductor Tunneling, Josephson Junctions and Squids

Consider a superconducting state

$$\Phi_0 = \prod_k (u_k + v_k \exp(i2\theta)\phi_k). \tag{6.70}$$

Then 2θ is called phase of superconductor.

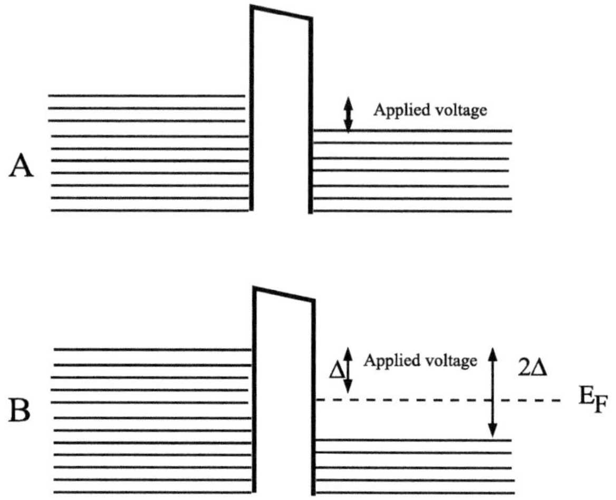

Fig. 6.10 Figure shows how an extra electron enters the superconductor with an energy Δ above the Fermi energy E_F. The applied voltage in **B** must then be $\frac{\Delta}{e}$ large for this tunneling to happen

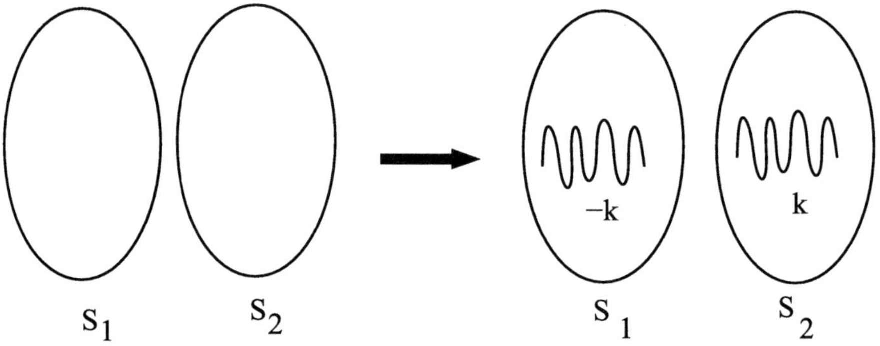

Fig. 6.11 Figure shows tunneling of a k wave from one superconductor to another creating excitation in both

Now consider a scenario, where we have two superconductors S_1 and S_2 as shown in Fig. 6.11 separated by a thin insulator called a weak link. This arrangement is called a Josephson junction. Then a electron with wave state $|k\rangle$ can tunnel from S_1 to S_2. This will cost energy as it will create an excitation both in S_1 and S_2 and raise the energy of the system by 2Δ. What is left behind in S_1 is state $|-k\rangle$. This process is energetically unfavorable, however immediately, $|-k\rangle$ can transition to S_2 making possible a second order process. This is shown in Fig. 6.12, where initial state of the system is two superconductors with n pair in each and we make a transition to intermediate state with wave state k in S_2 and $-k$ in S_1 and finally we transit to final state $n-1$ pairs in S_1 and $n+1$ pairs in S_2.

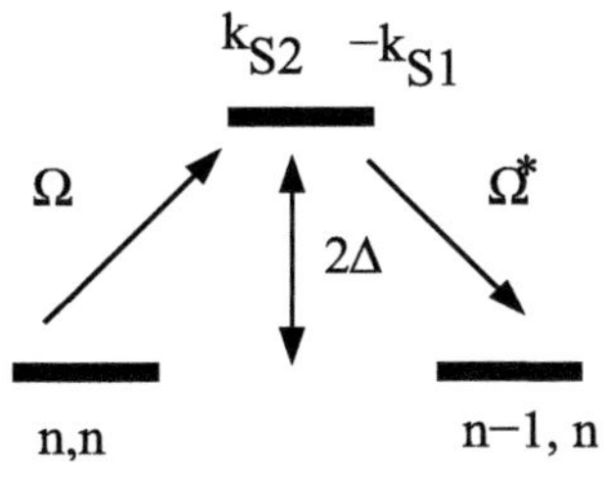

Fig. 6.12 Figure shows how tunneling of a Cooper pair between two superconductors can be represented as a three-level diagram, where intermediate state, level 2 is an excited state of both superconductors

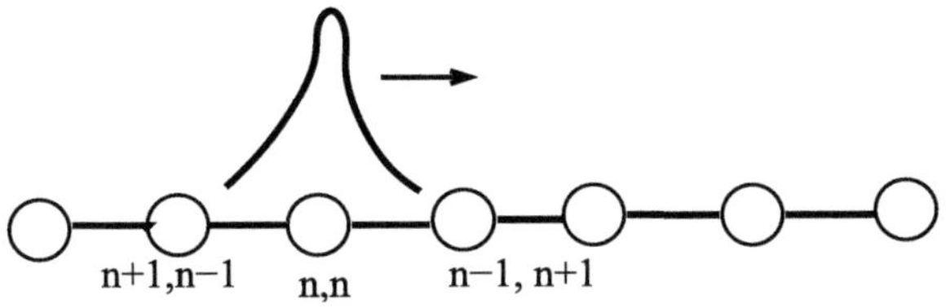

Fig. 6.13 Figure shows how tunneling of Cooper pairs can be represented as hopping between sites in a tight-binding model, where the lattice site represents a number of Cooper pairs in two superconductors. The initial state is represented as a wavepacket spread over a few lattice sites

If Ω is the transition rate to intermediate state, then the second order transition rate from initial to final state is

$$t = \frac{|\Omega|^2}{2\Delta} \tag{6.71}$$

Then we can describe the transition between the number states of the two superconductors as shown in Fig. 6.13 by a transition matrix

$$H = \begin{bmatrix} 0 & t & 0 & \dots & \dots & 0 \\ t & 0 & t & 0 & \ddots & \vdots \\ 0 & t & 0 & t & 0 & \vdots \\ \vdots & \ddots & \ddots & \ddots & \ddots & \vdots \\ 0 & \dots & \dots & t & 0 & t \\ 0 & 0 & \dots & \dots & t & 0 \end{bmatrix}. \tag{6.72}$$

If there is a phase difference 2θ between the superconductors this transition matrix takes the form

$$H_\theta = \begin{bmatrix} 0 & t\exp(i2\theta) & 0 & \dots & \dots & 0 \\ t\exp(-i2\theta) & 0 & t\exp(i2\theta) & 0 & \ddots & \vdots \\ 0 & t\exp(-i2\theta) & 0 & t\exp(i2\theta) & 0 & \vdots \\ \vdots & \ddots & \ddots & \ddots & \ddots & \vdots \\ 0 & \dots & \dots & t\exp(-i2\theta) & 0 & t\exp(i2\theta) \\ 0 & 0 & \dots & \dots & t\exp(-i2\theta) & 0 \end{bmatrix}. \quad (6.73)$$

Observe Hamiltonian in (6.72) is reminiscent of *tight-binding approximation.* How does a wavepacket $\phi(x)$ as shown in Fig. 6.13 evolve under H_θ as in (6.73). It evolves the same way as evolving $\exp(ikx)\phi(x)$ under H with $ka = 2\theta$, with a thought of as lattice spacing in 1D tight-binding model. But we know the energy for a tight-binding model is

$$\omega(k) = 2t\cos(ka). \quad (6.74)$$

Then the group velocity of the packet is s

$$v_g = 2at\cos(2\theta), \quad (6.75)$$

and therefore current in superconductor S_2 is

$$i = 2qv_g/a = 4qt\cos(2\theta). \quad (6.76)$$

If we apply a potential difference between the superconductors then $\frac{d\theta}{dt} = \frac{qV}{\hbar}$ and therefore the current

$$i = 2qv_g/a = 4qt\cos\left(\frac{2qVt}{\hbar}\right). \quad (6.77)$$

Now consider the local BCS states/superconductors $S_1, \dots, S_n$ as in Fig. 6.14 put in a loop. If we turn on a magnetic field (say in time T) through the center of the loop, it will establish a transient electric field in the loop given by

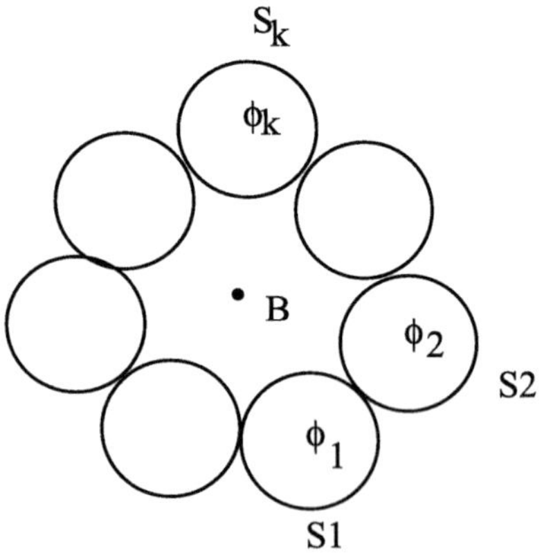

Fig. 6.14 Figure shows local BCS states/superconductors arranged in a loop, threaded by a magnetic field in the center, leading to the phase difference between adjacent superconductors

$$\int_0^T E(\tau) = \frac{Ba_r}{2\pi r} = \frac{\Phi_0}{2\pi r},$$

where r is radius and a_r area of the loop. The electric field in the loop means the potential difference between adjacent superconductors which gives superconductor phases ϕ_k, with $\Delta\phi$ as the phase difference between adjacent superconductors. If distance between adjacent superconductors is Δx, with $n\Delta x = 2\pi r$, then

$$\Delta\phi = \frac{(2e\Delta x)\int_0^T E(\tau)}{\hbar} = \frac{(2e\Delta x)\Phi_0}{2\pi r\hbar}.$$

Since we have a closed loop

$$n\Delta\phi = \frac{2e\Phi_0}{\hbar} = 2\pi, \tag{6.78}$$

giving

$$\Phi_0 = Ba_r = \frac{h}{2e}.$$

This is the magnetic quantum flux. When one deals with the superconducting loop or a hole in a bulk superconductor, it turns out that the magnetic flux threading such a hole/loop is quantized [4, 5] as just shown.

Figure 6.15 depicts the schematic of a superconducting quantum interference device (SQUID), where two superconductors S_1 and S_2 are separated by thin insulators. A small flux through the SQUID creates a phase difference in the two superconductors (see discussion on $\Delta\phi$ above) leading to the flow of super-current.

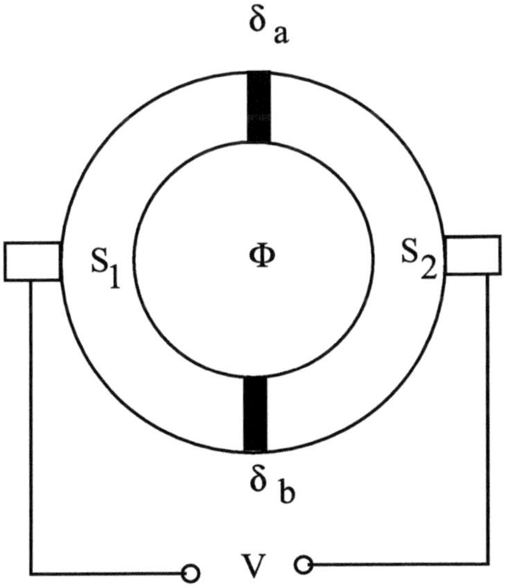

Fig. 6.15 Figure depicts the schematic of a SQUID where two superconductors S_1 and S_2 are separated by thin insulators

If an initial phase δ_0 exists between the superconductors. Then this phase difference after application of flux is from (6.78), $\delta_a = \delta_0 + \frac{e\Phi_0}{\hbar}$ across a top insulator, and $\delta_a = \delta_0 - \frac{e\Phi_0}{\hbar}$ across bottom insulator, see Fig. 6.15. This leads to currents $J_a = I_0 \sin \delta_a$ and $J_b = I_0 \sin \delta_b$ through top and down insulators. The total current $J = J_a + J_b = 2I_0 \sin \delta_0 \cos \frac{e\Phi_0}{\hbar}$. This accumulates charge on one side of SQUID and leads to a potential difference between the two superconductors. Therefore flux is converted to a voltage difference. The voltage oscillates as the phase difference $\frac{e\Phi_0}{\hbar}$ goes in integral multiples of π for every flux quanta Φ_0. SQUID is the most sensitive magnetic flux sensor currently known. The SQUID can be seen as a flux-to-voltage converter and it can generally be used to sense any quantity that can be transduced into a magnetic flux, such as electrical current, voltage, position, etc. The extreme sensitivity of the SQUID is utilized in many different fields of applications, including biomagnetism, materials science, metrology, astronomy, and geophysics.

Problems

1. Consider a 3D periodic potential

$$V = V_0 \cos^2\left(\frac{\pi x}{a}\right) \cos^2\left(\frac{\pi y}{a}\right) \cos^2\left(\frac{\pi z}{a}\right), \tag{6.79}$$

   with $V_0 = 10V$ and lattice parameter $a = 3A°$ amd atomic mass of 20 protons. Calculate the parameter c in the book and use it to find the binding energy per electron and superconducting gap assuming debye frequency of $\omega_d = 10^{13}$ Hz and Fermi energy $E_F = 10$ eV.
2. In the above problem if the hopping parameter $t = 5$ eV, calculate ω_F and binding energy and superconducting gap.
3. How will the superconducting gap change if the mass of the ions was doubled?
4. How will the superconducting gap change if the spring constant of the lattice was doubled?
5. How will the superconducting gap change if the lattice parameter a was doubled?
6. How will the superconducting gap change if the hopping parameter t was doubled?
7. What is minimum magnetic field allowed through a hole of radius 1 mm in a superconductor?

References

1. M. Tinkham, *Introduction to Superconductivity*, 2nd ed. (McGraw Hill, 1996)
2. P.G. De Gennes, *Superconductivity of Metals and Alloys* (Benjamin, Inc, W.A, 1966)
3. J. Bardeen, L. Cooper, JR Schriffer, Theory of superconductivity. Phys. Rev. **108**(5), 1175 (1957)
4. B. Deaver, W. Fairbank, William "experimental evidence for quantized flux in superconducting cylinders". Phys. Rev. Lett. **7**(2), 43–46 (1961)
5. R. Doll, M. Näbauer, Experimental proof of magnetic flux quantization in a superconducting ring. Phys. Rev. Lett. **7**(2), 51–52 (1961)

Chapter 7
Electrons in Magnetic Fields

In this chapter, we study electrons in magnetic field. Consider an electron moving in the 2D plane with magnetic field B applied in the z direction. Then the electron feels the Lorentz force evB, perpendicular to its motion, and its direction changes. for radius r such that

$$\frac{mv^2}{r} = evB \tag{7.1}$$

the electron executes a circular motion, called cyclotron orbit, with radius r and velocity $v = \omega r$, such that its angular velocity satisfies

$$\omega = \frac{eB}{m}. \tag{7.2}$$

For a field of $B = 10$ T, we get $\omega \sim 10^{12}$ rad/s.

Now consider a conductor in magnetic field along z direction, with current moving along x axis as shown in Fig. 7.1. The electron moving along x axis, feels a Lorentz force along y axis, which pushes them to edges and leads to development of a repulsive hall field E_H (hall voltage V_H) along y direction which balances this force, which gives

$$\frac{V_H}{d} = E_H = vB. \tag{7.3}$$

The velocity v is related to current density j_x as $j_x = nev$ where n is the carrier density and $j_x = I/A$, where A is the cross-sectional area of the conductor. Substituting we get

$$R_H = \frac{V_H}{I} = \frac{\Phi}{N_0 e}, \tag{7.4}$$

N. Khaneja, *Electron Waves in Solids*, Springer Series in Solid-State Sciences 203,
https://doi.org/10.1007/978-3-031-67260-6_7

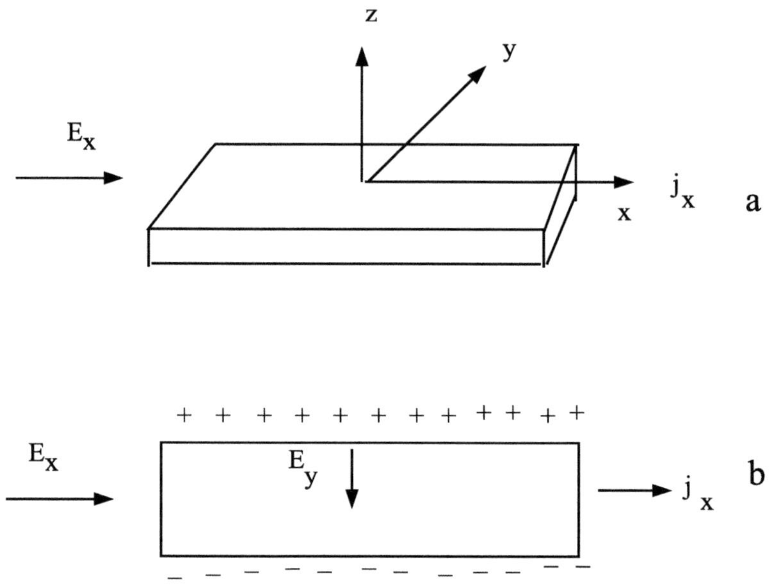

Fig. 7.1 Figure a shows a conductor in magnetic field along z direction with current moving along x axis. Fig, b shows the top view of the conductor with Hall field in the y direction

where N_0 is total carriers in the conductor, and Φ magnetic flux through the sample. This is called the classical Hall effect. When we put a current carrying conductor in the magnetic field, it develops a Hall voltage. The voltage develops because the electron moving along x direction is deflected along y direction. This will make charge accumulate along the top and bottom edge of the conductor as shown in Fig. 7.1b which gives the hall field E_H and voltage V_H.

On closer inspection, we may find that electrons in the bulk of the 2D plane will be deflected along y direction, and they will just execute cyclotron motion. However at edges, we do not have room to execute cyclotron motion. These electrons then press against the edge and develop a Hall field due to electron repulsion. Therefore one may conclude, that the Hall field is seen only be edge electrons and they give rise to Hall voltage the bulk does not. This phenomenon is revealed in a classical set of experiments called the quantum hall effect. We describe this effect in the next section.

7.1 Quantum Hall Effect

The quantum hall effect is a quantum-mechanical version of the Hall effect, observed in two-dimensional electron systems, subjected to strong magnetic fields and low temperatures, in which the Hall conductance σ undergoes quantum hall transitions to take on the quantized values [1, 2].

$$\sigma = \frac{I_{channel}}{V_{Hall}} = \nu \frac{e^2}{h}, \tag{7.5}$$

where $I_{channel}$ is the channel current and V_{hall} is the Hall voltage, e the electron charge and h Planck's constant The prefactor ν is called the filling factor and can take on either integer ($\nu = 1, 2, 3$) or fractional ($\nu = \frac{1}{3}, \frac{2}{5}$) values [1, 2]. The quantum hall effect is referred to as integer or fractional depending on if ν is an integer or a fraction.

The quantization of the Hall conductance has the important property of being exceedingly precise. Actual measurements of the Hall conductance have been found to be an integer or fractional multiples of $\frac{e^2}{h}$, to nearly one part in a billion. This has allowed for definition of a new practical standard for electrical resistance, based on the resistance quantum given by von Klitzing constant $R_K = \frac{h}{e^2} = 25812.807557\,\Omega$. This is named after Klaus von Klitzing who in 1980 made the unexpected discovery that the Hall conductivity was exactly quantized [1].

In this chapter, we study the plateaus that are characteristic of Hall resistance in the quantum hall effect. A two-dimensional (2D) electron system as found in the inversion layer of a MOSFET, when placed in strong magnetic field B, shows characteristic plateaus, when Hall resistance R_H is plotted as function of gate voltage V_g at fixed B, or as function of B at fixed V_g. In literature, we attribute this characteristic to disorder in the system and the presence of extended and localized states. Here we present an alternate theory that does not need disorder. Using a quasiparticle approach, we show that electrons in the bulk of such a 2D system do not carry any current. They simply execute a cyclotron motion. All current is carried by edge states, we call boundary layer, which is the source of the Hall voltage. When we increase gate voltage and fill in electrons, we first fill the bulk, and then the boundary layer and hence we have to wait before we increment the electrons in the boundary layer which gives the characteristic plateaus.

7.1.1 Theory

Consider a two-dimensional electron system as shown in Fig. 7.2 with magnetic field B perpendicular to the plane. The classical electron with velocity v in plane orbits around the field B, with centripetal force

$$\frac{mv^2}{r} = evB \tag{7.6}$$

with $v = \omega r$, the angular velocity $\omega = \frac{eB}{m}$ with radius of the cyclotron orbit $\frac{v}{\omega} = \frac{mv}{eB}$.

In Fig. 7.2, consider an electron moving from left to right with velocity v. It feels like an upward force evB which is offseted by a Hall field E_H pointing vertically up, which pushes the electron down and balances the force due to the magnetic field as $evB = eE_H$ where $E_H = \frac{V_H}{L_x}$, where V_H is Hall voltage, from which we get $V_H = vBL_x$. Now the current I which is left to right is $I = jL_xd$, where $j = n_cev$

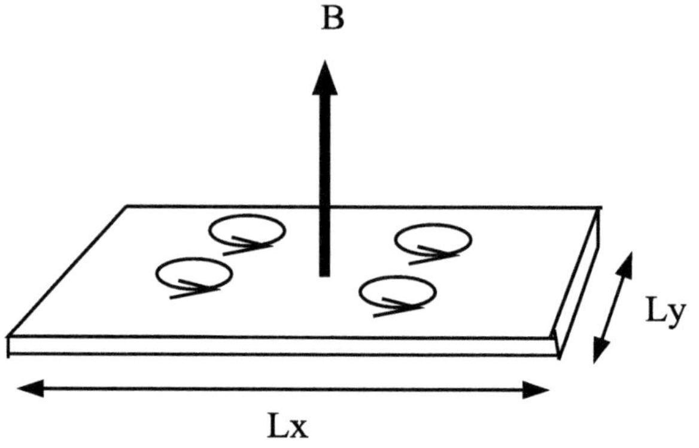

Fig. 7.2 Figure shows a two-dimensional electron system in magnetic field B

is the current density, n_c carrier concentration, v carrier velocity and d the thickness of the 2d plate. Then this gives $v = \frac{I}{n_c d L_x e}$ or $V_H = \frac{\Phi I}{e N_0}$, with N_0 total number of carriers and $\Phi = BL_x L_y$ the flux through the plane. The Hall resistance

$$R_H = \frac{V_H}{I} = \frac{\Phi}{eN_0}. \tag{7.7}$$

Let's calculate electron energies in the magnetic field B. We write the Schröedinger equation

$$i\hbar\frac{\partial \phi}{\partial t} = H\phi, \tag{7.8}$$

where H is the Hamiltonian of the system.

$$H = \frac{1}{2m}\{(-i\hbar\frac{\partial}{\partial x} + eA_x)^2 + (-i\hbar\frac{\partial}{\partial y} + eA_y)^2\}, \tag{7.9}$$

where A_x, A_y are vector potentials. We choose Landau Gauge with $A_x = 0$ and $A_y = Bx$. This gives

$$H = \frac{1}{2m}\{(-i\hbar\frac{\partial}{\partial x})^2 + (-i\hbar\frac{\partial}{\partial y} + eBx)^2\}. \tag{7.10}$$

Then $\exp(ik_y y)\phi^n_{k_y}(x)$ is an eigenfunction, where $\phi^n_{k_y}$ is a eigenfunction of the Harmonic oscillator Hamiltonian

$$H_0 = \frac{-\hbar^2}{2m}\frac{\partial^2}{\partial x^2} + \frac{k}{2}(x + \frac{\hbar k_y}{eB})^2 \tag{7.11}$$

with $k = \frac{e^2B^2}{m}$. $\phi^n_{k_y}$ has energy $(n + \frac{1}{2})\hbar\omega$ where $\omega = \sqrt{\frac{k}{m}} = \frac{eB}{m}$ and $\phi^n_{k_y}$ is centered at $x(k_y) = -\frac{\hbar k_y}{eB}$.

The $x - y$ dimension of the 2D electron system is L_x, L_y respectively as in Fig. 7.2. Then we take k_y at spacing of $\frac{2\pi}{L_y}$ to localize electrons in the y dimension. Then for each n, we have total number of modes (also called Landau levels) as

$$M_0 = L_x(\frac{\hbar 2\pi}{eBL_y})^{-1} = \frac{e\Phi}{2\pi\hbar},$$

where $\Phi = BL_xL_y$ is the total flux passing through the 2D system. For a sample measuring $100's$ of microns on sides and $B = 10$ tesla we find $M_0 \sim 10^7$.

Observe total electrons in the plane $N_0 = \nu M_0$, where ν are different $n's$ in the harmonic oscillator. Then substituting in (7.7) we get

$$R_H = \frac{\Phi}{e\nu M_0} = \frac{2\pi\hbar}{e^2\nu}. \tag{7.12}$$

Recall N_0 denotes the total number of electrons. As we increase N_0, the M_0 landau levels get filled corresponding to $n = 0$. Then again M_0 landau levels get filled corresponding to $n = 1$ and so on. Thus ν increases from 1 to higher values.

Conversely for fixed N_0, for large B, number of Landau levels M_0 are large and all electrons can be accommodated in $n = 0$. As we decrease B, M_0 decreases and for same N_0, ν increases.

A two-dimensional electron system is realized [1] as an inversion layer in a MOSFET, with n-doped source and drain and a p channel connecting them. A positive bias on gate voltage pulls the electrons near the gate-channel interface and creates a 2D electron system. N_0 can be controlled by changing gate voltage V_g and the whole setup is in a magnetic field B. The direction from source to drain is our y direction and transverse to it is the x direction. When we measure Hall resistance for a fixed B (say at 10 T), and variable N_0, by changing V_g, we find a characteristic as in Fig. 7.3A. When we fix V_g, and vary B, we get a characteristic as in Fig. 7.3B.

Now we come to the main point of this chapter. Why are there characteristic plateaus? In literature, this is attributed to disorder and the presence of localized and extended states [2]. The localized states do not carry current and all current is carried by extended states. In this paper, we present an alternate theory that does not need disorder. On application of the magnetic field, electrons reside in Landau levels. The Landau electrons don't evolve in space as the different k_x have the same energy then a k_x wavepacket doesn't travel. This is shown as cyclotrons in the bulk in Fig. 7.4. But if we apply an electric field along x axis the Landau state increases in $k_x \rightarrow k_x + \frac{qEt}{\hbar}$, which pushes the Landau particle to the top of the plate. This is because the wavepacket $\exp(k_x x)\phi_n(y - y_0)$ is centered at $y_0 = \frac{\hbar k_x}{qB}$, as k_x changes the packet center moves.

We still need to show as $y_0(t) = \frac{\hbar k_x(t)}{qB}$ changes, the process is adiabatic and we stay in the eigenstate, $\exp(k_x x)\phi_n(y - y_0(t))$. For $n = 0$, we have $\phi_0 \propto \frac{1}{sqrt\sigma} exp(-\frac{(y-y_0(t))^2}{2\sigma^2})$, then

$$\hbar\langle\dot{\phi_0}(t), \phi_1\rangle = \frac{\hbar E}{B\sigma},$$

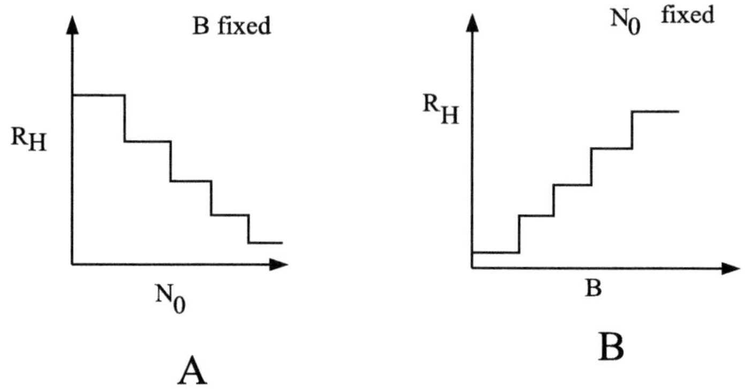

Fig. 7.3 Figure A shows a characteristic plot of Hall resistance R_H for a fixed B (say at 10 T) and variable N_0 When we fix V_g, and vary B, we get a characteristic as in fig. B. Increments of R_H are in units of $\frac{h}{e^2}$ while B is in tens of Tesla and N_0 increments are $\sim 10^7$

where $\sigma^2 = \frac{\hbar}{m\omega_0}$, for adiabaticity $\frac{\hbar E}{B\sigma} \ll \hbar\omega_0$, the energy spacing between the eigenvalues. For $E \sim 100\text{V/m}$ and $B \sim 10\text{T}$, we have the adiabaticity condition read $\frac{m}{\hbar} < 10^{11}$, which is true hence we are in adiabatic regime and electric field drags the packet upwards on the plane.

More landau states accumulate at the top, which we call edge states. But the edge states have their harmonic potential bottom y_0 outside the plane, but they themselves are not outside hence they gain in energy and more energy for larger k, which means the edge state will travel, and this is the current, only at the top of the plane, shown as arcs in Fig. 7.4. Accumulated edge states give the Hall voltage. We can write Landau states as x states $\exp(k_x x)\phi_n(y - y_0)$ or as y states $\exp(k_y y)\phi_n(x - x_0)$. When hall voltage becomes big, will push the y landau states to the side (move x_0) opposing the applied field till it is nulled.

Now we are ready to answer plateaus. The formula in (7.7) and (7.12) are true in the boundary layer because it is here we have x direction current. The bulk just doesn't evolve. Therefore N_0, ν all make sense for the boundary layer. The electrons in top boundary layer in Fig. 7.4 travel from left to right. Application of electric field (voltage between drain and source) gives electron crowd and vacuum at the top and bottom respectively, resulting in Hall voltage. Now for a fixed B, as we fill electrons, we fill the bulk first and then the boundary layer as shown in Fig. 7.5. As we fill, ν in the boundary layer increments. We have to wait to fill the bulk first and then the boundary layer increments. This gives the plateaus and we get discrete jumps in Fig. 7.3A. Similarly, when we fix N_0 and change B, we are changing M_0 and hence for large B, first Landau levels can accommodate all electrons. As we decrease B, we increase ν again first filling bulk and then filling boundary layer and with decreasing B, we have characteristic as in Fig. 7.3B with the plateaus.

In summary, plateaus arise because bulk does not carry any current. All current is in the boundary layer the edge states.

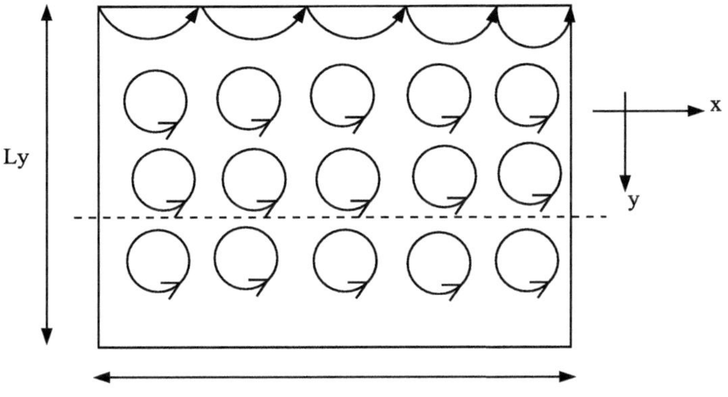

Fig. 7.4 Figure shows how bulk of 2D system executes cyclotron motion. Current is carried by edge states

Fig. 7.5 Figure shows how we fill Landau levels, first bulk then edges as they have slightly higher energy

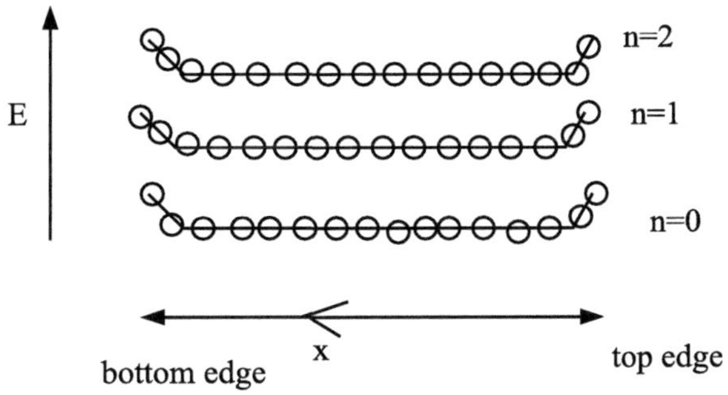

7.2 Fractional Quantum Hall Effect

In the quantum hall effect, we have $\frac{e\Phi}{h}$ degenerate Landau level that fills the 2D plane. At large magnetic fields, this may even exceed the total particles N_0. Then how do we fill? We may not even have anything at the edge, and that means very high hall resistance. However, there is repulsion between the electrons. Which means even if say $N_0 = \frac{1}{3}$ of $(\frac{e\Phi}{h})$, we may spread these electrons on the plane, so that we minimize the repulsion. Which means we will still get edge states though now at fractional filling $\nu = \frac{1}{3}$. This is the basis of the fractional quantum hall effect. We observe plateaus in hall resistance as a function of the magnetic field, which corresponds to fractional filling ν. The plateaus have been seen for filling fractions $\nu = \frac{1}{5}, \frac{2}{5}, \frac{1}{3}, \frac{2}{3}, \frac{3}{7}, \frac{4}{9}$ etc. [3].

Problems

1. In classical hall effect if conductor has top face area cm^2 and thickness mm, find the Hall voltage for electron density $10^{29}/\text{m}^3$ and $B = 1$ T and current $I = 1$ A.
2. For magnetic field $B = 10\text{T}$, find the cyclotron frequency.
3. For a plane $100 \times 100\,\mu$, find the landau degeneracy at magnetic field of 10T.
4. In above find a number of Landau levels if a number of electrons $N_0 = 10^{10}$.
5. In above find the Hall resistance and filling factor.

References

1. K.V. Klitzing, G. Dorda, M. Pepper, Phys. Rev. Lett. **45**, 494 (1980)
2. D. Tong, "Quantum Hall Effect", TIFR Infosys Letures (2016). http://www.damtp.cam.ac.uk/user/tong/qhe.html
3. D.C. Tsui, H.L. Stormer, A.C. Gossard, Phys. Rev. Lett. **48**, 1559 (1982)

Chapter 8
Magnetism

In this chapter we study the fundamental solid state phenomenon of magnetism [1]. Magnetism has many manifestations, like ferromagnetism, paramagnetism, diamagnetism, antiferromagnetism, etc.

Ferromagnetism is the basic mechanism, by which certain materials (such as iron) form permanent magnets, or are attracted to magnets. An everyday example of ferromagnetism is a refrigerator magnet, used to hold notes on a refrigerator door. Permanent magnets (materials that can be magnetized by an external magnetic field and remain magnetized after the external field is removed) are either ferromagnetic or ferrimagnetic, as are the materials that are noticeably attracted to them. Only a few substances are ferromagnetic. The common ones are iron, nickel, cobalt, and most of their alloys, and some compounds of rare earth metals. Ferromagnetism is very important in industry and modern technology and is the basis for many electrical and electromechanical devices, such as electromagnets, electric motors, generators, transformers, and magnetic storage such as tape recorders, and hard disks, and nondestructive testing of ferrous materials.

Solid materials may lose or gain energy in the magnetic field and hence are attracted or repelled by the magnetic field and are called paramagnetic and diamagnetic respectively. The applied magnetic field H_0 induces a magnetic moment $M = \chi H_0$ per unit volume. The total $H = (1 + \chi)H_0$ and

$$B = \mu_0 H = \mu_0(1 + \chi)H_0. \tag{8.1}$$

Then χ is called magnetic susceptibility, which is positive for paramagnetic and negative for diamagnetic. The induced field in the paramagnet supports, while in a diamagnet, opposes the applied field. Metals like silver, gold, lead, zinc, copper, bismuth, and mercury are all diamagnetic. Their susceptibility is negative.

Paramagnetism arises as excess spins get oriented in the direction of the magnetic field. Diamagnetism arises as electron waves develop a magnetic moment that is

N. Khaneja, *Electron Waves in Solids*, Springer Series in Solid-State Sciences 203,
https://doi.org/10.1007/978-3-031-67260-6_8

oriented opposite to the applied field. In ferromagnetism, we saw how spins spontaneously align in the same direction. Antiferromagnetism is a kind of magnetic ordering in which spins of neighboring atomic sites are anti-align.

8.1 Diamagnetism

In the last section, we talked about paramagnetic metals. Metals like silver, gold, lead, zinc, copper, bismuth, and mercury are all diamagnetic. Their susceptibility is negative. To understand diamagnetic susceptibility we have to understand the diamagnetism of free electrons also called *Landau Diamagnetism.*

8.1.1 Diamagnetism of Free Electrons: Landau Diamagnetism

Free electrons also contribute to diamagnetism. Let's calculate the susceptibility. When we apply a magnetic field, it induces a magnetic moment M, such that

$$\mu_0 \frac{M}{V} = -\chi B, \tag{8.2}$$

where χ is susceptibility. The energy in a magnetic field is $E = -M \cdot B = \frac{V}{\mu_0} \chi B^2$ and hence

$$\chi = \frac{\mu_0}{2V} \frac{\partial^2 E}{\partial B^2}, \tag{8.3}$$

where V is total volume.

To calculate energy E we need eigenfunctions in the magnetic field B in z direction as in Fig. 8.1. The eigenfunction as shown in last chapter is of the form $\psi = \exp(ik_z z)\exp(ik_x x)\phi_n(y)$, with energy $\frac{\hbar^2 k_z^2}{2m} + E_n$ where $\phi_n(y)$, E_n satisfies the equation

$$\left(\frac{-1}{2m} \frac{\partial^2}{\partial y^2} + \frac{m\omega_0^2}{2} (\frac{\hbar k_x}{qB} + y) \right) \phi_n = E_n \phi_n. \tag{8.4}$$

Clearly ϕ_n are Harmonic oscillator eigenfunctions with energy $E_n = (n + \frac{1}{2})\hbar\omega_0$ with $\omega_0 = \frac{qB}{m}$. k_x steps in increments of $\frac{2\pi}{L_x}$ and hence number of eigenfunctions for given n is $\Delta = \frac{qLxLyB}{h} = \frac{q\Phi_0}{h}$, where Φ_0 is flux through plane area in Fig. 8.1.

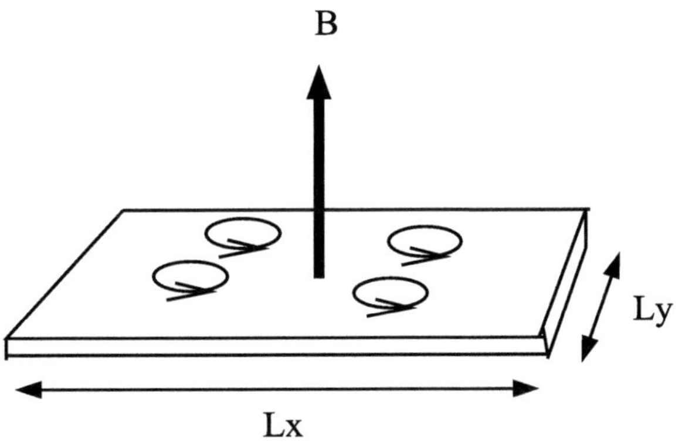

Fig. 8.1 Figure shows a two-dimensional electron system in magnetic field B

Then the total energy is from n filled Landau levels and $n+1$ filled to fraction ξ such that total electrons $N = (n+\xi)\Delta$

$$E = \frac{L_z}{a}\left(\Delta\hbar\omega_0\left(\sum_j (j-\frac{1}{2}) + (n+\frac{1}{2})\xi\right)\right), \tag{8.5}$$

where $\frac{L_z}{a}$ counts number of planes, where a is lattice spacing.

Then

$$\chi = \frac{\mu_0}{2V}\frac{\partial^2 E}{\partial B^2} = \frac{\mu_0 q^2}{4\pi m a}\xi(1-\xi). \tag{8.6}$$

For $a \sim 3A^\circ$, we have $\chi \sim 10$ ppm (10^{-6}). If there are Z electrons per atom, we have

$$\chi = \frac{Z\mu_0 q^2}{4\pi m a}\xi(1-\xi). \tag{8.7}$$

When B changes ξ changes from 0 to 1 and χ oscillates. This phenomenon of oscillation of χ is called **De Haas–van Alphen effect**.

The de Haas–van Alphen effect is useful in measuring the fermi surface of metals. As we vary the magnetic field, we find the susceptibility oscillates. This is the de Haas–van Alphen effect. Consider a metal in a magnetic field along say z direction. Figure 8.2 a shows the fermi surface of the metal. Shown are extremal areas A_1 and A_2 at specific k_z values. There are many electrons at the fermi surface for these specific k_z values. Let's take one of these extremal areas say A_2, call A_e.

At successive oscillations, with field strength B_1, B_2, we have Landau degeneracy Δ_1 and Δ_2 such that

$$\Delta_1 n = \Delta_2(n+1) = N \tag{8.8}$$

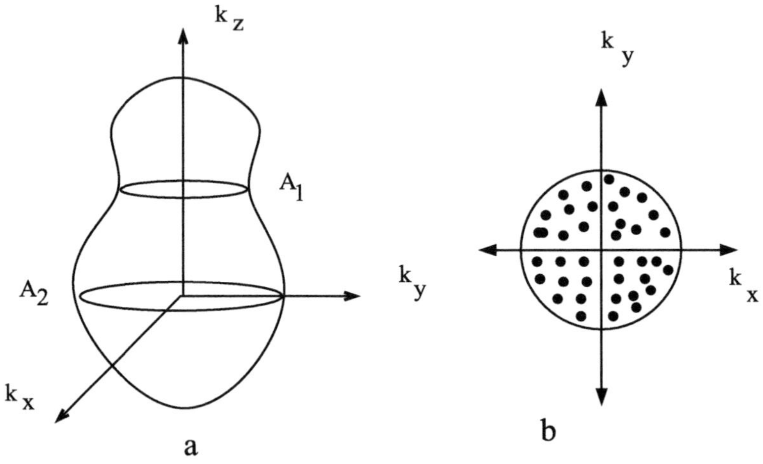

Fig. 8.2 Figure shows the fermi surface for a metal with extremal areas A_1 and A_2

$$A_e(\frac{1}{B_2} - \frac{1}{B_1}) = \frac{2\pi e}{\hbar}. \tag{8.9}$$

From the above equation, we can find A_e. Hence we can find the extremal area of Fermi surface by a period of this oscillation and by choosing different directions for the magnetic field we can map the Fermi surface.

8.2 Paramagnetism

Solid materials may lose or gain energy in the magnetic field and hence are attracted or repelled by the magnetic field and are called paramagnetic and diamagnetic, respectively. The applied magnetic field H_0 induces a magnetic moment $M = \chi H_0$ per unit volume. The total $H = (1 + \chi)H_0$ and

$$B = \mu_0 H = \mu_0(1 + \chi)H_0. \tag{8.10}$$

Then χ is called magnetic susceptibility which is positive for paramagnetic and negative for diamagnetic. The induced field in the paramagnet supports, while in a diamagnet, opposes the applied field.

Metals with partially filled shells have equal electrons in the spin up and down state as shown in Fig. 8.3a. On application of the magnetic field, the energy of spin up state is lowered by an amount $\mu \cdot B$, where $\mu = \frac{\hbar q}{2}$ m is magnetic moment of spin, and energy of the spin down state is raised by an amount $\mu \cdot B$. This prompts more electrons to move to spin up state, lowering the overall energy as shown in

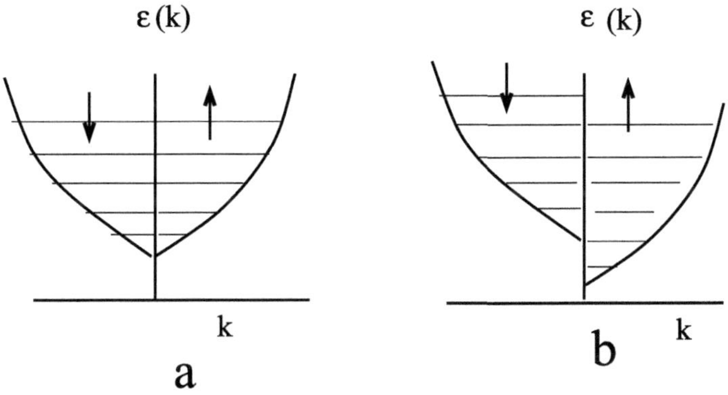

Fig. 8.3 Figure shows how in the magnetic field, energy of states with spin up is lowered and spin down is raised

Fig. 8.3b. This is called *Pauli-paramagnetism.* Lithium, sodium, magnesium, aluminum, cesium, etc. are some examples of paramagnetic metals. Most transition metals including scandium, titanium, vanadium, chromium, manganese, molybdenum, etc., are all paramagnetic. Similarly, most of rare earth metals like cerium, praseodymium, neodymium, samarium, europium, etc., are all paramagnetic.

If at Fermi level with energy ϵ_f, a small band of energy $\Delta E = \mu B$ of down spins is converted to up. The number of states in this small band $= D(\epsilon_f)\Delta E$ and the resulting magnetic moment is $\mu D(\epsilon_f)\Delta E$. But if we consider a Fermi sphere of radius k_f. Then in a small annulus Δk we have number of electrons ΔN such that $\frac{\Delta N}{N} = \frac{3\Delta k}{k_f}$, where N is number of electrons in fermi sphere. Furthermore Fermi energy $\epsilon_f = \frac{\hbar^2 k_f^2}{2m}$ which gives $\frac{\Delta\epsilon}{\epsilon_f} = \frac{2\Delta k}{k_f}$, which gives $D(\epsilon_f) = \frac{3N}{2\epsilon_f}$. Then the resulting magnetic moment per unit volume is $M = \frac{3n\mu^2 B}{2\epsilon_f}$, where n is density and

$$\chi = \frac{dM}{dH} = \frac{3n\mu_0\mu^2}{2\epsilon_f}. \tag{8.11}$$

This is the paramagnetic susceptibility of free electrons.

8.3 Exchange and Ferrmomagnetism

Consider two electrons with wavevector k_1 and k_2, then they don't know which one is k_1 and k_2 and the wavefunction is a symmetrized one

$$\psi(r_1, r_2) = \frac{1}{\sqrt{2}}\left(\frac{\exp(ik_1 \cdot r_1)\exp(ik_2 \cdot r_2) + \exp(ik_2 \cdot r_1)\exp(ik_1 \cdot r_2)}{\sqrt{2}}\right).$$

The total energy for $k = k_1 - k_2$ is

$$E_T = \int \psi^\dagger(r_1, r_2) \frac{1}{4\pi\epsilon_0(r_1 - r_2)} \psi(r_1, r_2) dr_1 dr_2 \tag{8.12}$$

$$= E_{coulomb} + E_{exchange} \tag{8.13}$$

$$E_{coulomb} = \int \frac{1}{4\pi\epsilon_0(r_1 - r_2)} dr_1 dr_2 \tag{8.14}$$

$$E_{exchange} = \int \frac{\cos(k \cdot (r_1 - r_2)}{4\pi\epsilon_0(r_1 - r_2)} dr_1 dr_2. \tag{8.15}$$

If $k = 0$, the two electrons are same state and hence have large exchange energy (Pauli exclusion principle). If $kl \gg 1$, where l is volume dimension, then exchange integral oscillates fast and close to 0. In general if electrons are on in wave orbitals ϕ_a and ϕ_b, we have

$$E_{coulomb} = \int \frac{|\phi_a(r_1)|^2 |\phi_b(r_2)|^2}{4\pi\epsilon_0(r_1 - r_2)} dr_1 dr_2 \tag{8.16}$$

$$E_{exchange} = \int \frac{\phi_a^*(r_2)\phi_b^*(r_1)\phi_a(r_1)\phi_b(r_2)}{4\pi\epsilon_0(r_1 - r_2)} dr_1 dr_2. \tag{8.17}$$

Id ϕ_a, ϕ_b disjoint $E_{exchange} = 0$. Of course exchange energy is 0 if we have unlike spins.

Consider conduction electrons in a metal. Imagine we have half filled bands as shown in Fig. 8.4a. Then we have as many spin up as down. Figure 8.4b shows a ferromagnetic metal with excess of spin up compared to spin down. Spin up is full band.

Why we have this excess of spins and full band can be understood when we look at exchange repulsion between electrons. When we have full band, we can localize them to atomic sites and hence make exchange interaction between like spins negligible. The band under consideration is usually made out of d-orbitals so-called d-band. This is because d-bands have small bandwidth, i.e., t parameter is small, and hence kinetic energy is small. Then making the band full and localization doesn't cost much. This is called Ferromagnetism, where excess spins get localized on atomic sites as in metals iron, nickel, and cobalt.

Thus ferromagnetism arises in an attempt to minimize exchange repulsion which is or order of eV. When we have more spins of one kind as in Fig. 8.4b, we have to pay a price as we are populating k-states with higher energy. When the conduction band has a small bandwidth as a d-band formed from d-orbitals, then this price is less severe. This is the case in iron, nickel, cobalt, etc. Metals like sodium, calcium, and aluminum that are primarily s band have large bandwidths and are therefore not ferromagnetic.

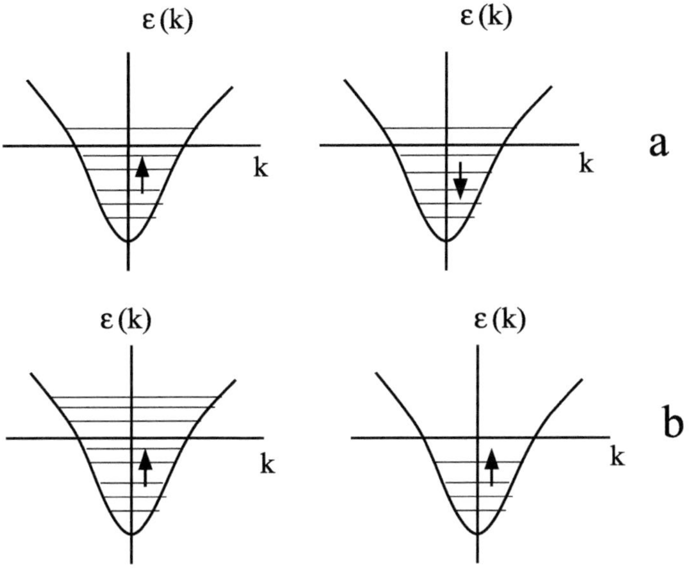

Fig. 8.4 **a** Shows half filled bands in a metal, with as many spin up as down. **b** Shows a ferromagnetic metal with excess of spin up compared to spin down

8.4 Antiferromagnetism

In ferromagnetism, we saw how spins spontaneously align in the same direction. Antiferromagnetism is a kind of magnetic ordering in which spins of neighboring atomic sites are anti-align. Instead of electrons in bands, we are talking now about electrons on atomic sites. When a band is full, we can form a superposition of k-states, such a superposition is localized to an atomic site. We can make a replica of this and put it at all atomic sites. These states also called *Wannier states* are localized and their total energy is the same as the sum total of energy of all k states. Except Wannier states do not overlap and hence they have minimum electron–electron repulsion between them.

When we have a narrow d-band which is half filled with two electrons in every k-state, it makes sense to fill all k-states with one electron per state instead and then *Wannierize* them. This way we save the energy of electron repulsion. Of course you pay a price in filling high energy k-states but that is not much for narrow bandwidth d-bands. This is shown in Fig. 8.5. It is in this sense we talk about atomic sites. We have one electron in all k-states and after *Wannierization*, we have one electron per site. This phenomenon occurs commonly among transition metal compounds, especially oxides. Examples include hematite, metals such as chromium, alloys such as iron manganese (FeMn), and oxides such as nickel oxide (NiO). Transition metals have unpaired electrons in d-orbitals which form narrow d-bands.

Now why anti-align neighboring sites? This opens the possibility that one can hop to a neighboring site that has an opposite spin. If you recall, our discussion of

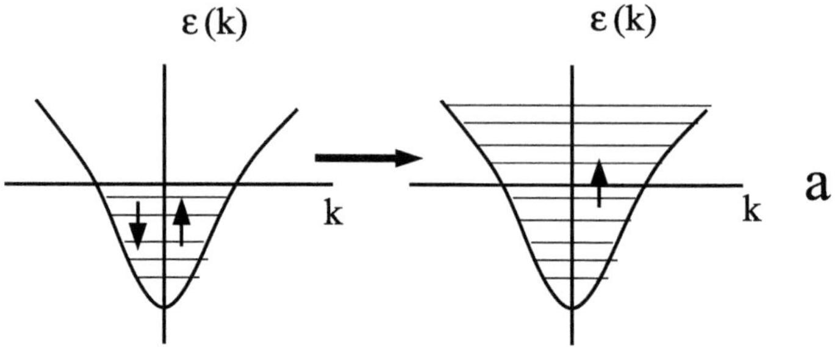

Fig. 8.5 Figure shows how a mott-insulator is formed by filling all k-states

bounding-antibonding orbital, this hopping is done through a transfer term of strength $-t$. If onsite energy is ϵ_0, then the energy of state A with electron on neighboring site is $2\epsilon_0$ and energy of state B with two electrons on same site is $2\epsilon_0 + U$, where U is the electron–electron repulsion. Then Hamiltonian for state A and B takes the form

$$H_0 = \begin{bmatrix} 2\epsilon_0 & -t \\ -t & 2\epsilon_0 + U \end{bmatrix}. \tag{8.18}$$

On diagonalization we find that energy $2\epsilon_0 \rightarrow 2\epsilon_0 - \frac{t^2}{U}$. Anti-alignment helps to reduce the energy. This is how one may explain antiferromagnetism. When we Wannierize, we automatically fill all k states making a fully filled band which is an insulator. Such an insulator where we have a narrow d-band and we localize electrons to save in electron repulsion is called *Mott-Insulator*.

Problems

1. For lattice spacing $a = 3$ A°, find the diamagnetic susceptibility for $Z = 1$.
2. In above find paramagnetic susceptibility for $\epsilon_f = 5$ eV and $n = 10^{28}/\text{m}^3$.
3. In above find paramagnetic susceptibility if hopping parameter $t = 2$ eV and lattice parameter is $a = 3$ A°.
4. Find exchange energy between two like spin electrons with wavevector difference $\Delta k = 10^8/\text{m}$.
5. For repulsion energy $U = 5$ eV and hopping parameter $t = 2$ eV, find the energy reduction due to antiferromagnetic ordering.

Reference

1. S. Blundell, Magnetism in Condensed Matter. Oxford University Press (2001)

Chapter 9
Imaging Solids

9.1 X-Ray Diffraction

X-ray diffraction is an important modality, for determining crystal structures [1–3]. Inter-atomic spacing in a crystal is few A°. The wavelength of X-ray light is of the same order. When light travels between atoms, it gathers a phase, which is the product of inter-atomic distance and the wavenumber, and hence for large wavenumbers as in X-rays, this gathered phase is sizeable and can give information about inter-atomic spacing by wave interference.

9.1.1 X-Ray Scattering and Antenna Arrays

Atoms scatter light (X-ray light in this case). Each atom acts as an antenna, resulting from induced dipole due to incident light, and acts as a current element that radiates. The collection of atoms acts as an antenna array. The array is displayed as in Fig. 9.1. Observe the atoms, as we move horizontally right, and get the radiation later, hence the induced current in the antenna is phase delayed. However if the incident and reflected wave make the same angle θ with the vertical, the path traveled by atoms on the right is shorter by the same amount as the delay, and the two factors cancel to give a coherent addition of reflected radiation from the atoms in a horizontal layer. How do different layers add coherently? As shown in Fig. 9.1, the path difference between the reflected light by successive layers is $2d \sin\theta$ and for the wavefronts to coherently add up, we should have for wavenumber $k = \frac{2\pi}{\lambda}$, where λ is the wavelength,

$$2k\, d \sin\theta = 2n\pi, \tag{9.1}$$

N. Khaneja, *Electron Waves in Solids*, Springer Series in Solid-State Sciences 203,
https://doi.org/10.1007/978-3-031-67260-6_9

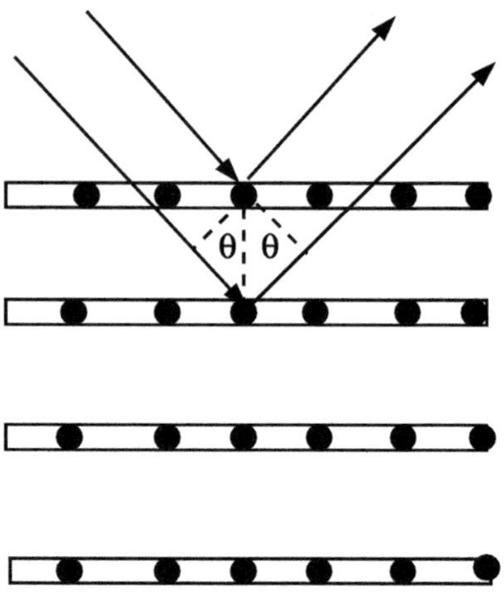

Fig. 9.1 Figure shows atomic layers that reflect X-rays in a crystal

or

$$2d \sin \theta = n\lambda. \tag{9.2}$$

This condition is called *Bragg's condition.*

Then if θ_1 is the angle at which we get coherent addition and if we change θ (rotate our crystal) till we again get coherent addition at θ_2, then we have

$$2d(\sin \theta_2 - \sin \theta_1) = \lambda \tag{9.3}$$

from which we can determine d the spacing between lattice planes. This is how we can use X-rays to determine the structure of the material.

9.2 Neutron Diffraction

Thermal electrons have wavelengths of 50 A°. Neutrons are 10^3 times as massive, and hence have a wavelength of a couple of A°. Then like X-rays, they can be used for finding crystal structures [4]. Now instead of X-ray light, we have neutrons. The question then is how does a neutron wave, scatter of atomic lattice. The neutron wave, when it passes by the nucleus of the atom will occupy the same space as the nucleons. Fermions cannot be at the same place and hence it will see a repulsive potential from the nucleus. Scattering of neutron arises due to this repulsive potential.

As shown in Fig. 9.2 the incoming neutron wave with wavevector k_1, scatters of a nucleus to k_2, with a scattering matrix element $V_{k_1,k_2} = \langle \exp(ik_2 \cdot r)|V|\exp(ik_1 \cdot r)\rangle$. Then the transition element from another nucleus that is displaced by vector r_0 is $\exp(i(k_1 - k_2) \cdot r_0)V_{k_1,k_2}$. In general, the scattering amplitude due to all the nuclei is then

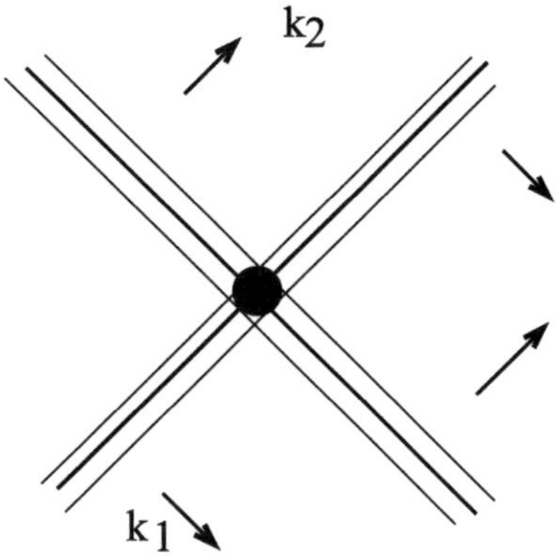

Fig. 9.2 Figure shows scattering of neutron plane wave with momentum k_1 to k_2

$$V_{k_1,k_2} \sum_n \exp(i(k_1 - k_2) \cdot r_n). \tag{9.4}$$

For all the amplitudes to add coherently, it should be the case that

$$(k_1 - k_2) \cdot r_0 = 2n\pi, \tag{9.5}$$

where r_0 is a lattice displacement vector that displaces you from one lattice point to neighboring lattice point. This condition is called *Laue's Condition.*

In Fig. 9.3, when k_1 and k_2, are as shown making an angle θ with the horizontal. Then observe

$$k_1 - k_2 = \frac{4\pi \sin\theta}{\lambda}. \tag{9.6}$$

along $-z$ direction. Then (9.5), implies that

$$2d \sin\theta = n\lambda \tag{9.7}$$

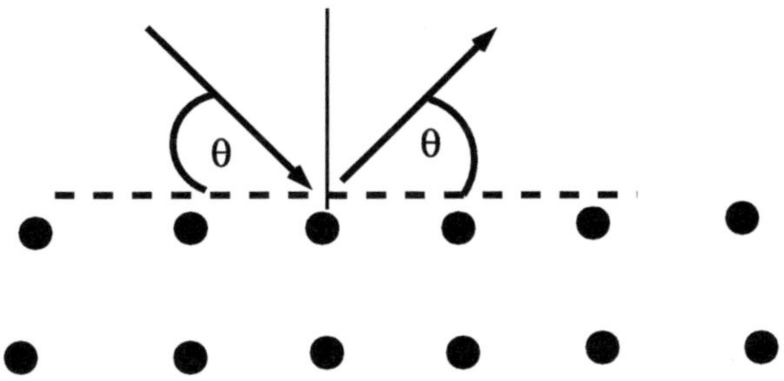

Fig. 9.3 Figure shows neutron wave scattering of atomic nuclei

as seen before in X-ray diffraction, we have recovered Bragg's condition. For correct θ, as in (9.7), we find a detector in direction of k_2 registers scattered neutrons.

9.3 Neutron Spectroscopy of Phonons

In neutron diffraction, we studied the scattering of neutron waves from stationary targets of the atomic nuclei. In neutron spectroscopy [5], we study the scattering of neutron waves from oscillating atomic nuclei, the phonons (lattice deformation waves). The neutron inside a crystal feels lattice potential and exchanges momentum with the lattice. In particular, a neutron with momentum k_i, when interacts with a phonon of momentum k changes its momentum to $k_f = k_i + k$ and in the process the phonon is annihilated. Conservation of energy dictates that if $\hbar\omega$ is the phonon energy then

$$\frac{\hbar^2 k_f^2}{2m} - \frac{\hbar^2 k_i^2}{2m} = \hbar\omega. \tag{9.8}$$

By detecting the momentum and energy of the scattered neutron, we can find the momentum and energy of the phonon and hence its dispersion relation. Figure 9.4 shows a schematic of a neutron spectrometer. Incident beam of neutrons are passed through a monochromator to choose to choose for a particular incident momentum/wavelength. The neutron after scattering from the sample is passed through an analyzer crystal to read its momentum. Both monochromator and analyzer work on the principle of neutron/Bragg diffraction. Figure 9.5 shows the energy/momentum relation of the annihilated phonon can be used to measure the

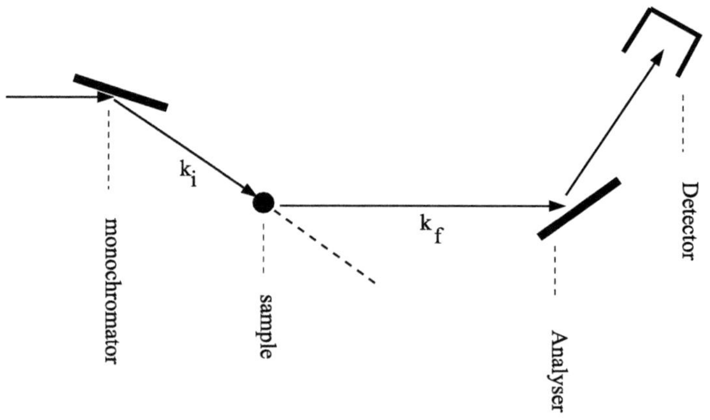

Fig. 9.4 Figure shows schematic of a neutron spectrometer

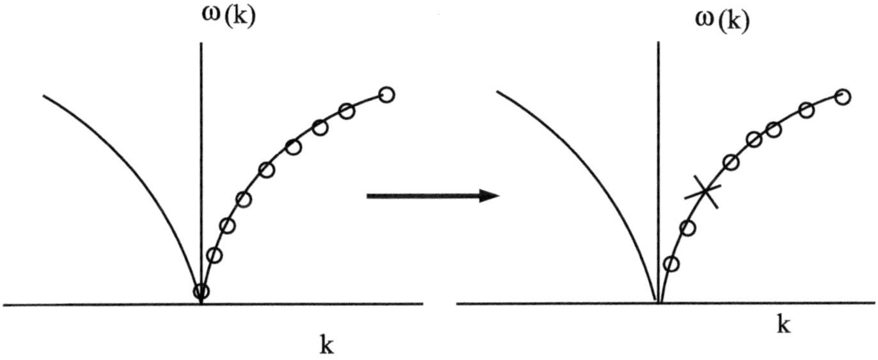

Fig. 9.5 Figure shows phonon dispersion curve and phonons at different, $k, \omega(k)$ pairs. After scattering with a neutron, the phonons lose energy, and momentum at certain $k, \omega(k)$ location

phonon dispersion curve. By changing the orientation of the analyzer we can sample different k_f on the dispersion curve.

9.4 Angle Resolved Photoelectron Spectroscopy (ARPES)

An important modality for band mapping in solids is angle resolved photoelectron spectroscopy (ARPES) [6].

Incident photons of the right energy (X-ray), when impinging the surface of a solid, eject valence electrons in the direction given by the wavevector of electron (momentum of photon is insignificant compared to electron momentum). Putting a detector in that direction and finding the energy of the ejected electron and using the incident photon energy photon can generate $(k, \omega(k))$ information, and map the band in a solid. ARPES is also called microscope into momentum space of valence electrons.

Consider free Bloch electrons, with energy organized as energy band. Incoming photon promotes the Bloch electron in the valence band to higher energy band, as shown in Fig. 9.7. The Bloch states in higher energy band are almost plane wave states. The higher energy band has positive total energy, and electron in this band is able to escape the lattice and reach the detector. The transition from valence band to higher band preserves energy. Momentum is also preserved as the incoming photon has very small momentum, compared to electron. We call the momentum of the higher energy band with plane wave-like states as k_i, with i standing from the in-crystal state. This plane wave leaves the lattice and becomes a free electron and we call its momentum k. The energy of k_i is

$$E_i = \frac{\hbar^2 k_i^2}{2\,m} - V_0, \tag{9.9}$$

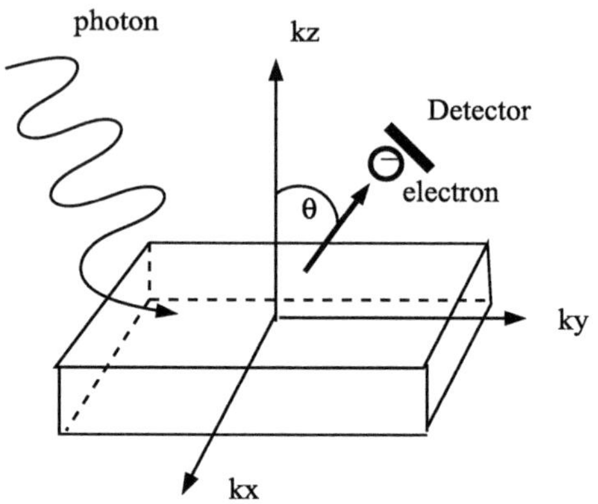

Fig. 9.6 Figure shows the basic setup of angle resolved photoelectron spectroscopy. Incident photons of the right energy when impinging the surface of a solid, eject valence electrons in the direction given by the wavevector of the electron, which are detected by a detector at the right angular position and analyzed for their energy

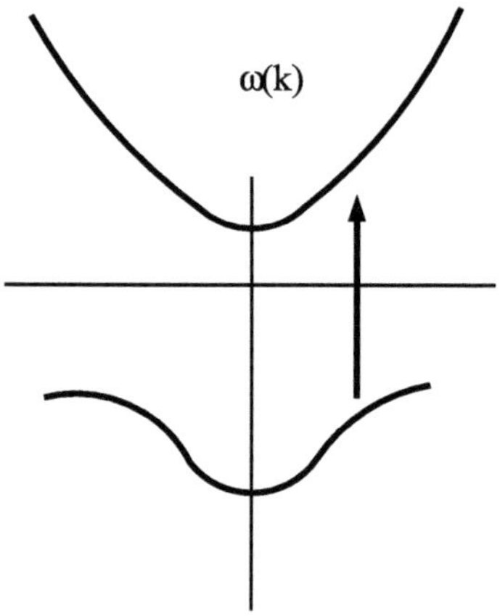

Fig. 9.7 Figure shows the transition from lower energy band to higher energy band in photo electron spectroscopy

where V_0 is average lattice potential as in (4.15) (we take V_0 positive here and put negative sign outside). Then in-crystal wave as it moves out follows Snell's law of refraction, i.e., $(k_i)_\parallel = (k)_\parallel$ (wave inside and outside the crystal is matched at the interface). Where $\parallel$ is the momentum parallel to the surface plane as in Fig. 9.6. Using $E_i = E$, we get from

$$\frac{\hbar^2 (k_i)_\perp^2}{2m} - V_0 = \frac{\hbar^2 k_\perp^2}{2m} = E\cos\theta. \tag{9.10}$$

$$\frac{\hbar^2 (k_i)_\parallel^2}{2m} = \frac{\hbar^2 k_\parallel^2}{2m} = E\sin\theta. \tag{9.11}$$

From (E, θ), we can construct k_i the band momentum and from $E_i - \hbar\omega$, where $\hbar\omega$ is the photon energy we can know the band energy and this way we can map the band.

9.5 Electron Microscopy

How do we see small? Small objects with small separations between them can be seen if we use light with small wavelengths. Then the light will bounce off the objects, but photons will also pass through the gaps between objects, and illuminate a screen behind the sample. Small wavelength means transmitted photons do not diffract and do not illuminate all the screens behind them. The transmitted light will then carry dark spots of objects, that reflect electrons, and bright spots of gaps between them. If a light photon has a big wavelength it will diffract as it passes through the gaps, and we will only see a big bright spot.

In electron microscopy [7], instead of light we use electron bars/wavepackets to see small. The de Broglie wavelength at room temperature is calculated as

$$\frac{1}{2}mv^2 = \frac{3}{2}kT, \tag{9.12}$$

which gives for $\hbar k = mv$ that $k^{-1} = \lambda = \frac{\hbar}{\sqrt{3kTm}}$, which gives, $\lambda = 1nm$.

In practice, we accelerate electrons to energies of the order of 100 kev, where we now have de Broglie wavelengths in the sub-angstrom regime. If such small electrons are fired at a sample, there are parts of the sample that scatter them and do not let them pass through but the gaps between such parts can be as small as angstrom, and still the small electrons will pass through these pores without much diffraction and hence give a very high resolution image where we can see gaps between scatterers as bright spots and scatterers as dark spots. This is the principle of transmission electron microscopy.

Electron microscopes are used to investigate the ultra-structure of a wide range of biological and inorganic specimens including microorganisms, cells, large molecules, biopsy samples, metals, and crystals. Industrially, electron microscopes are often used for quality control and failure analysis.

Another mode of operation of electron microscopes is called scanning electron microscope, whereby electrons are focused on a spot of nanometer resolution, and reflected electrons are a measure of contrast of the spot. Different objects give different contrasts generating an image of nanometer resolution.

9.6 Scanning Tunneling Microscopy

In this section, we study an important technology for imaging the surfaces of solids, the Scanning Tunneling Microscopy [8].

In March 1981 G. Binnig, H. Rohrer, Ch. Gerber and E. Weibel at the IBM Zurich Research Laboratory observed the vacuum tunneling of electrons between a sharp tungsten tip and a platinum sample. Combined with the ability to scan the tip against the sample surface, the scanning tunneling microscope (STM) was born. Since then,

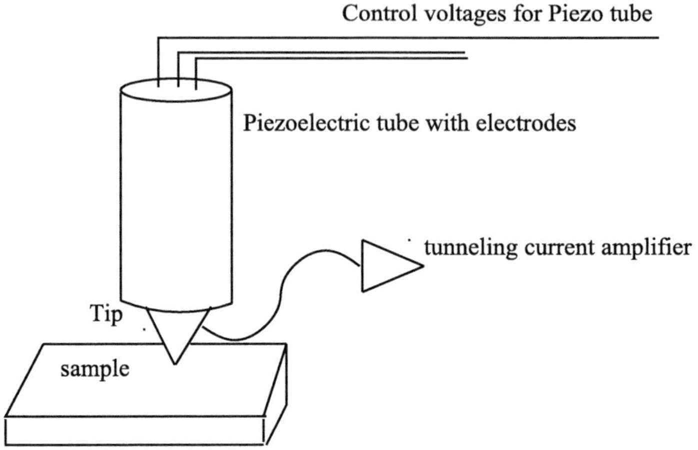

Fig. 9.8 Figure shows a schematic of STM

this novel type of microscopy has continuously broadened our perception of atomic scale structures and processes. The STM allows one to image atomic structures directly in real space, giving us the opportunity to make the beauty of nature at the atomic level directly visible. Moreover, the sharp tip can be regarded as a powerful local probe that allows one to measure the physical properties of materials on a small scale by using a variety of different spectroscopic methods (Fig. 9.8).

The general principle of operation of a scanning tunneling microscope (STM) and related scanning probe microscopies (SPM) as well is surprisingly simple. In STM a bias voltage is applied between a sharp metal tip and a conducting sample to be investigated (metal or doped semiconductor). After bringing the tip and sample surface within a separation of only a few Angstrom, a tunneling current can flow due to the quantum mechanical tunneling effect before 'mechanical point contact' between tip and sample is reached. The tunneling current can be used to probe physical properties locally at the sample surface as well as to control the separation between tip and sample surface. The distance control based on tunneling is very sensitive to small changes in separation between the two electrodes because the tunneling current is strongly (exponentially) dependent on this separation, as we will see later. By scanning the tip over the sample surface while keeping the tunneling current constant by means of a feedback loop, we can follow the surface contours with the tip which to a first approximation will remain at constant distance from the sample surface. By monitoring the vertical position z of the tip as a function of the lateral position (x, y), we can get a three-dimensional image z(x,y) of the sample surface. Motion of the tip both laterally and vertically with respect to the sample surface can be realized with sub-atomic accuracy by means of piezoelectric drives.

In contrast to other electron microscopes and surface analytical techniques using electrons, STM can be operated in air and in liquids as well as in a vacuum because there are no free electrons involved in the STM experiment. Therefore, the application of STM is not limited to surface science, but has particularly great potential for in situ electro- chemical studies and in vivo investigations of biological specimens.

The brightest prospects for STM/SPM are offered in the field of nanometer-scale science and technology. As electronic devices become increasingly smaller, there is a strong need for understanding the physical properties of matter on a nanometer scale. Technology at the nano- meter level requires nanopositioning and control, nanoprecision machining, and reproducible creation of nanometer-scale structures as well as the use and control of super-smooth surfaces.

Since tunneling is at the heart of STM technology, we review the basics of tunneling. In the following, we revisit the classical problem of an electron tunneling through a potential barrier. The classical treatment on the subject considers an electron wave at energy E, and potential barrier of height U with $E < U$. One computes the transmission and reflection coefficients for this wave by solving for the eigenfunction of the Schröedinger equation in different regions and matching the solution at region boundaries. We review the classical treatment on tunneling. In Sect. 9.6.2, we provide a wavepacket interpretation of tunneling. We show how we make sense of incident, transmitted, and reflected waves, i.e., how do we calculate the dynamics of wavepacket that is incident on the barrier? In this section, we show how the classical solution can be understood in terms of wavepackets and resulting tunneling dynamics computed.

9.6.1 Wave Solution

Consider the potential barrier of height U as shown in Fig. 9.9. The solution with energy $E < U$ to the Schröedinger equation,

$$i\hbar\frac{\partial\phi}{\partial t} = -\frac{\hbar^2}{2m}\frac{\partial^2\phi}{\partial x^2} + V(x)\phi \tag{9.13}$$

for this potential can be written in region I as

$$\phi = \exp(ikx) + R\exp(-ikx), \tag{9.14}$$

where $E = \frac{\hbar^2 k^2}{2m}$. The solution in region II is

$$\phi = A\exp(k_1 x) + B\exp(-k_1 x), \tag{9.15}$$

where $U - E = \frac{\hbar^2 k_1^2}{2m}$. Finally solution in region III is

$$\phi = T\exp(ikx). \tag{9.16}$$

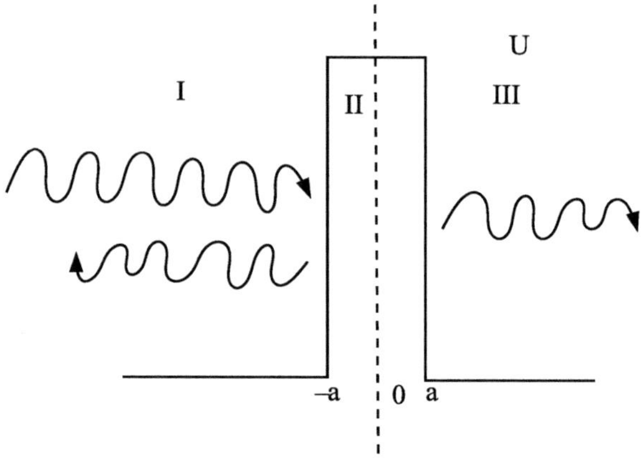

Fig. 9.9 Figure shows a potential barrier and incident, reflected and transmitted wave

We call R and T, reflection and transmission coefficients. To determine R and T, we match ϕ and ϕ' at $x = -a$ and $x = a$ in Fig. 9.9. This gives

$$\begin{aligned}
\exp(-ika) + R\exp(ika) &= A\exp(-k_1 a) + B\exp(k_1 a).\\
ik(\exp(-ika) - R\exp(ika)) &= k_1(A\exp(-k_1 a) - B\exp(k_1 a)).\\
T\exp(ika) &= A\exp(k_1 a) + B\exp(-k_1 a).\\
ikT\exp(ika) &= k_1(A\exp(k_1 a) - B\exp(-k_1 a)).
\end{aligned}$$

We can solve these equations to get

$$T = \frac{\exp(-i2ka)}{\cosh 2k_1 a - \frac{c+c^{-1}}{2}\sinh 2k_1 a}, \tag{9.17}$$

where $c = \frac{-ik_1}{k}$, with

$$|T| = \frac{1}{\sqrt{\cosh^2 2k_1 a + \frac{2+\frac{k^2}{k_1^2}+\frac{k_1^2}{k^2}}{4}\sinh^2 2k_1 a}}. \tag{9.18}$$

$$|R| = \frac{2(1+\frac{k_1^2}{k^2})\sinh 2a}{\sqrt{\cosh^2 2k_1 a + \frac{2+\frac{k^2}{k_1^2}+\frac{k_1^2}{k^2}}{4}\sinh^2 2k_1 a}}. \tag{9.19}$$

We call $T\exp(ikx)$ is region III and $R\exp(-ikx)$ in region I as transmitted and reflected wave. However, all we really have is ϕ, the eigenstate of the system Hamiltonian. We are interpreting parts of this wavefunction as incident, reflected, and transmitted wave. How can we justify this by genuinely talking about an incident electron wavepacket and calculating how it gets transmitted and reflected? We study this in the following section.

9.6.2 Wavepacket Interpretation

In the previous section, we wrote a wavefunction ϕ. Let's call it ϕ_k, as it is parameterized by the wavenumber k. Let the corresponding energy be E_k. We can write ϕ_k as the sum of three parts (Fig. 9.10)

$$\phi_k = \phi_k^a + \phi_k^b + \phi_k^c, \tag{9.20}$$

where $\phi_k^a = \exp(ikx),\ x < 0$ is the incident wave, $\phi_k^b = R\exp(-ikx),\ x < 0$ the reflected wave and $\phi_k^c = T\exp(-ikx),\ x > 0$, the transmitted wave.

Now consider the superposition

$$\psi = \sum \exp(ikx) \tag{9.21}$$

of different k centered around a nominal k_0. This is shown on k, E_k dispersion curve Fig. 9.11A.

The superposition takes the form

$$\begin{aligned}\psi &= \exp(ik_0x)\sum_k \exp(i(k-k_0)x) = 2\exp(ik_0x)\sum_{n=0}^{N}\cos n\Delta kx \\ &= 2\exp(ik_0x)\frac{B}{\Delta k}\frac{\sin(Bx)}{Bx} = \exp(ik_0x)f(x),\end{aligned}$$

where we take $k's$, Δk apart, with total bandwidth $B = N\Delta k$. $f(x)$ is a sinc function as shown in Fig. 9.11B.

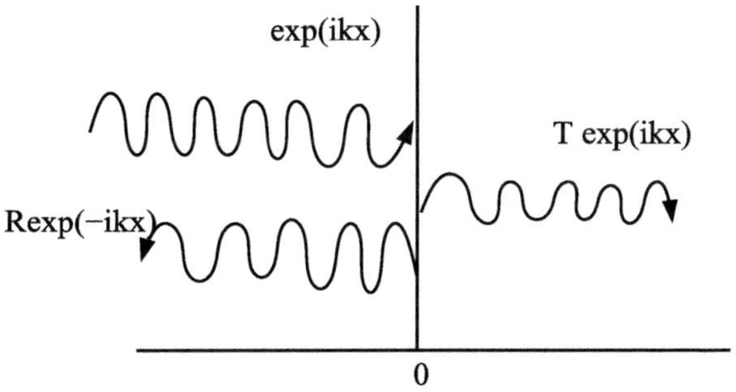

Fig. 9.10 Figure shows decomposition of an eigenstate as incident, reflected, and transmitted wave

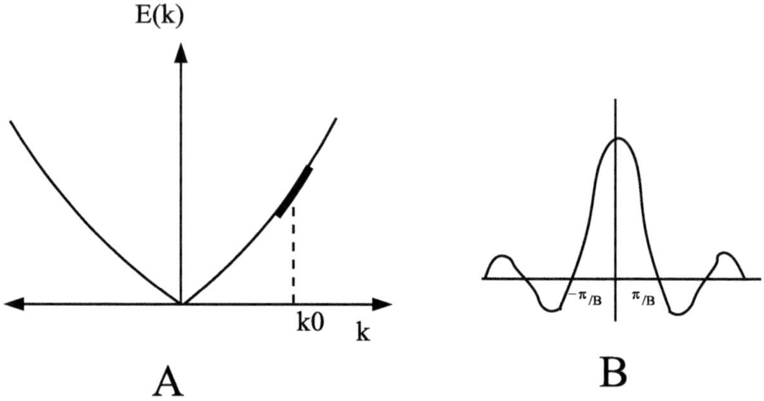

Fig. 9.11 **A** Shows the dispersion curve for free electron wave, with a packet prepared around k_0. **B** Shows such a wavepacket

How does this wavepacket evolve?

$$\begin{aligned}\psi(x,t) &= \exp(ik_0x)\sum_k \exp(-i\omega_k t)\exp(i(k-k_0)x)\\ &= \exp(ik_0x)\exp(-i\omega_0 t)\sum_{n=0}^{N}\exp(-i\omega'(k_0)(k-k_0)t)\exp(i(k-k_0)x)\\ &= \exp(ik_0x)f(x-\omega'(k_0)t),\end{aligned}$$

where $v_g = \omega'(k_0) = \frac{d\omega_k}{dk}|_{k_0}$ is the group velocity of the packet. The wavepacket travels from left to right with velocity v_g.

Now let's evolve the packet $\psi(x)$ for a negative time starting from the origin at time 0. Then at $-T$, the packet is far left of the origin at $x = -b$. Therefore at negative time $-T$ we have coefficients $a_k(-T) = \exp(i\omega_k T)$ such that

$$\sum_k a_k(-T)\phi_k = \sum_k a_k(-T)(\phi_k^a + \phi_k^b + \phi_k^c) = \sum_k a_k(-T)\phi_k^a = \psi(x+b),$$

where for these coefficients $a_k(-T)$, the contribution of ϕ_k^b is zero as for $-k$, this superposition gives a packet on the right hand of origin (mirror of one for positive k) but there is no ϕ_k^b on right. Similarly, the contribution of ϕ_k^c is zero as packet for positive k is produced on the left-hand side of origin but ϕ_k^c exits on right side. Now when we let this superposition evolve as $\sum_k a_k(-T)\exp(-i\omega_k t)\phi_k$, then the packet travels right till it hits the origin, after which positive k produce a packet on right and negative k on left and hence we get contribution from ϕ_k^b as a reflected packet

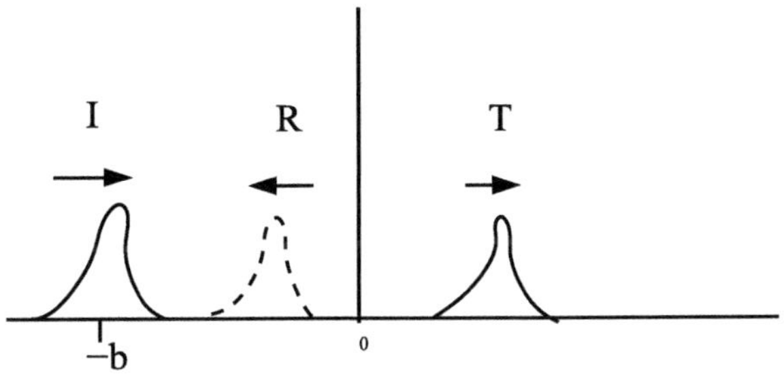

Fig. 9.12 Figure shows incident, reflected, and transmitted packet

on the left and contributions from ϕ_k^c as a transmitted packet on the right and no contribution from ϕ_k^a. All this is depicted in Fig. 9.12.

All we are saying is that an incident packet from the left can be written as a superposition of eigenstates ϕ_k and we know how this superposition evolves. It evolves into a reflected packet and a transmitted packet.

In summary in STM, a positive voltage on the tip pulls the electrons on the surface which then tunnel through the air gap. We reviewed the basics of tunneling above.

Problems

1. For a lattice with spacing $a = 3$ A° find the longest wavelength λ such that we get X-ray scattering at angle $\theta = 30°$.
2. Find the minimum energy of neutrons to get a neutron diffraction at an angle $\theta = 30°$.
3. In photoelectron spectroscopy, if work function $U = 3$ eV and kinetic energy of emitted electron is 1 eV, what is the wavelength of incident light?
4. If crystal is thought of as a square well with potential $U = -5$ eV. What is the angle and wavelength of refracted electrons when they are incident at an angle 30° with the normal with energy 10 eV.
5. A 1 eV electron is incident on a barrier with height 5 V and thickness 100 nm. What is the tunneling probability?

References

1. M.A. Glazer, The Structure of Crystals, Bristol (1987)
2. C. Hammond, The Basics of Crystallography and Diffraction. Oxford University Press (1997)
3. M.T. Dove, Structure and Dynamics. Oxford University Press (2003)
4. S.W. Lovesey, Theory of Neutron Scattering from Condensed Matter, vol. 1 and 2. Clarendon Press (1984)
5. G. Shirane, S.M. Shapiro, J.M. Tranquada, Neutron Scattering with a Triple-Axis Spectrometer: Basic Techniques. Cambridge University Press (2002)

6. F. Reinert, S. Hüfner, Photoemission spectroscopy from early days to recent applications. New J. Phys. **7**, 97 (2005)
7. S. Wistchnitzer, Introduction to Electron Microscopy. Pergamon Press (1970)
8. R. Wiesendanger, Scanning probe microscopy and spectroscopy: Methods and Applications. Cambridge University Press (1994)